Journeyman Electrician Exam Prep

Obtain Your License on the First Try Without Stress

Henry Bloom

Journeyman Electrician Exam Prep

TABLE OF CONTENTS

INTRODUCTION

How This Guide Can Help You Pass the Journeyman Electrician Exam

This comprehensive guide provides everything you need to pass the journeyman electrician exam. Through focused study strategies, hands-on examples, and an extensive collection of practice queries complete with clarifications, you'll acquire the expertise and self-assurance needed to excel in your examination.

By reviewing the detailed content in this guide, you'll develop a solid understanding of electrical fundamentals, system design and sizing, NEC regulations, and troubleshooting techniques.

The exam overview gives you insight into the structure and content of the test so you know what to expect. Efficient study tips help you make the most of your preparation time.

This guide includes the latest NEC updates so your knowledge reflects the current code. From branch circuit design to conductor selection, you'll be up-to-date on electrical best practices and safety standards. With a sharpened understanding of system sizing, motor controls, and kitchen appliance circuits, you'll confidently tackle any exam question.

Through using this all-encompassing prep guide, you can walk into the journeyman electrician exam fully prepared and equipped to pass. The targeted content review ensures you have a solid grasp of the most important concepts tested. Hundreds of practice questions allow you to apply your knowledge while identifying any weak areas needing more study. Detailed explanations provide the clarity you need to reinforce understanding.

With the tools and resources in this definitive guide, you can approach the journeyman electrician exam with confidence. The depth of material reviewed will arm you with the knowledge to pass with flying colors. So rely on this prep guide, study diligently, and get ready to ace the exam!

Overview of the Exam: Format, Content, and Scoring

Achieving success in this exam is pivotal for your professional progression, and being well-prepared with a thorough grasp of the exam's structure will empower you to tackle it confidently. This chapter provides an overview of the exam format, the content covered, and how the exam is scored.

The journeyman electrician exam is administered by your state or local licensing board. While the specifics vary slightly by jurisdiction, the exams follow a similar overall format. You can expect the exam to contain 100-150 multiple-choice questions testing your knowledge of electrical theory, code requirements, wiring techniques, and more. Most exams require to be completed within a 4-5 hour time limit. Questions will have four answer choices, and you must select the best answer. Some jurisdictions allow you to bring in code books, while others require you to work from memory.

The exam covers topics ranging from basic electrical concepts to complex system designs and troubleshooting. Key areas assessed include:

- Fundamentals of electricity - Ohm's Law, currents, voltages, power
- Electrical circuits - series, parallel, combinations, resistors
- Wiring techniques - sizing conductors, calculating loads, installation
- Overcurrent protection - fuses, circuit breakers
- Distribution equipment - transformers, regulators, capacitors
- Motors and motor circuits
- Lighting circuits and controls
- Appliance circuits - ranges, HVAC systems, water heaters
- Safety protocols and procedures
- NEC requirements for installations

You will need to demonstrate both broad knowledge and specific code applications. Expect scenario-based questions that test your judgment and problem-solving skills. Questions may provide diagrams, schematics, or descriptions of installations and ask you to determine code compliance, troubleshoot issues, or select appropriate materials and procedures.

The exam covers practical knowledge that an experienced electrician must possess. While you do need to study and prepare, the goal is to evaluate competency, not memorization skills. Keep this in mind as you get ready for the exam.

Scoring varies by jurisdiction but typically requires a 70-75% score to pass. Most licensing boards use computerized, adaptive exams that change the difficulty level based on your answers. This helps assess knowledge thoroughly. Results are given immediately after completion. You may have the option to retake the exam if needed.

Preparing for this exam requires dedication and practice. Understand the format, content, and scoring to guide your study plan. With a commitment to learning the material and sharpening skills, you have a high chance of passing on your first attempt. Use this book as a trusted resource along the way. The knowledge you gain will serve you well throughout your electrical career.

Effective Study Strategies and Exam Preparation Tips

Preparing for the Journeyman Electrician exam requires dedicating ample time to review key concepts and practice solving problems. By refining their study techniques, candidates preparing for their journeyman certification can approach the high-stakes examination with assurance.

To begin, compile a set of comprehensive study materials that align precisely with exam subject areas. Crucial resources include NEC code handbooks, magazines like EC&M that detail real-world applications, textbooks covering electrical theory and calculations, flashcards for formulas and definitions, and practice tests that mimic actual exam conditions. The closer the study tools match the real exam, the better.

With materials in hand, analyze practice tests to identify your weakest knowledge gaps. Allocate extra time to master concepts you struggle with. Study challenging topics first when your mind is fresh, not at the end when mental fatigue sets in. Connect new information to existing knowledge you already possess to aid retention. You should break study sessions into focused 25-minute blocks separated by brief breaks to maintain sharp focus.

Implement proven techniques like mnemonic devices to aid memorization. For example, the acronym OHM can help recall the key electrical quantities of Ohms, current in Amperes, and Voltage. POWER reminds us that Power equals Voltage multiplied by Current. Rely on vivid imagery and associations to link concepts. Visualize a giant Ohm's Law triangle crushing a house to symbolize the critical importance of understanding that formula.

Maximize retention through spaced repetition. Review material over several weeks rather than cramming the night before. Initially study a concept thoroughly, then circle back to it after a few days. Expanding the time between review sessions transfers knowledge into long-term memory. This prevents forgetting complex material as the exam date nears.

Stay organized by maintaining comprehensive notes as you study. Writing information down improves understanding compared to typing. Keep notes well-structured with topics clearly defined. Regularly update your notebooks and review them frequently. Consistent review of key facts, formulas, safety rules, and code references will prepare you for exam day.

Incorporate hands-on learning like building mock circuits on a breadboard or operating electrical equipment. Physical demonstration of concepts leads to true mastery. If possible, find a study partner to quiz and discuss concepts with. Teaching others requires crystallizing your understanding. When stumped, revisit foundational principles until the light bulb clicks.

Practice mock exams under real-world conditions. Time yourself and refrain from referencing materials as you take each test. By doing this, you can expose knowledge gaps to refine in the final weeks. Strive for the full exam duration without breaks to simulate real testing conditions. Maintain focus and avoid burnout with regular breathing exercises. Analyze mistakes thoroughly after each practice test.

In the final preparation stage, ramp up practice exam intensity. Take at least one full-length test weekly and identify areas needing improvement. Drill weak points relentlessly in the remaining weeks while maintaining knowledge in other areas through routine review. As the exam nears, get plenty of sleep and stick to normal routines. Light exercise can relieve stress while keeping the mind sharp.

When exam day arrives, stay confident with proper preparation. Eat a healthy meal and hydrate well to power your brain. Arrive early with ID, pencils, approved calculator, and other essentials. Listen to instructions carefully and read each question thoroughly before responding. Eliminate choices confidently and guess logically when unsure, since there is no penalty. Trust your abilities and remain focused to demonstrate comprehensive electrical knowledge. With diligent preparation and skills-focused study strategies, you will pass the Journeyman Exam with flying colors. Stay positive and go conquer that test!

CHAPTER 1
FUNDAMENTALS OF ELECTRICITY

Basic Concepts: Charge, Current, Voltage, and Power

Understanding the fundamentals of electricity starts with grasping some key concepts - charge, current, voltage, and power. Mastering these buildings prepares you for the foundation to tackle more complex electrical engineering principles.
Charge refers to the property bestowed upon the subatomic particles of matter - negatively charged electrons and positively charged protons. Current is the flow of electric charge; it measures the rate at which charge moves past a point over a certain period. It is measured in amperes. Voltage, measured in volts, is the electric potential energy required to move charge between two points in a circuit.
These parameters are linked by Ohm's law, which states that current is directly proportional to voltage and inversely proportional to resistance. Using Ohm's law, you can derive important relationships like:
Voltage = Current x Resistance
Power (watts) = Current (amps) x Voltage (volts)
Understanding these basic equations allows you to calculate unknown variables and determine how components like resistors influence current and voltage in a circuit.
Taking a deeper dive into current and voltage, alternating current (AC) refers to a charge that changes direction periodically. Direct current (DC) flows in one constant direction. Most household appliances run on 120V or 240V AC power.
Another important principle is Kirchhoff's Current Law, which states that the currents entering and exiting a node must sum to zero. Kirchhoff's Voltage Law states that the voltage drops around a closed loop must equal zero. Analyzing circuits using Kirchhoff's laws provides experience for the types of multi-loop troubleshooting problems you will encounter on the journeyman exam.
Gaining fluency with fundamental terminology and units of measurement establishes a critical foundation. You must internalize concepts like charge, current, voltage, resistance, and power to develop the intricate knowledge required to pass the journeyman exam and excel as an electrician.

Deep Dive into Ohm's Law

Ohm's law forms the foundation of electrical circuit analysis. Understanding it thoroughly is crucial for success as an electrician. This chapter explores Ohm's law in depth, moving from basic concepts to real-world applications.
To begin with, Ohm's law states that current is directly proportional to voltage, assuming resistance remains constant. This can be summarized by the equation: $V = IR$, where V is voltage measured in volts, I is current measured in amps, and R is resistance measured in ohms. While simple, this equation allows us to understand the relationship between these key circuit parameters.
For example, if a circuit has a voltage of 120V and a resistance of 10 ohms, using Ohm's law we can calculate that the current is 12 amps. Rearranging the equation also lets us calculate voltage or resistance if we know the other two values. This is immensely useful for designing and troubleshooting circuits.
We can extend Ohm's law to determine the power dissipated in a circuit. Power is measured in watts. Using $P = IV$, if a circuit draws 5 amps at 24 volts, the power can be calculated as 120 watts. This helps size wires, fuses, and other components.
Next, let's look at applications of Ohm's law in series and parallel circuits. With series circuits, the current remains constant while voltage drops across each resistor on the circuit. The total voltage equals the sum of the individual voltage drops. For parallel circuits, voltage remains constant while current is divided between branches. Using Ohm's law for each resistor allows us to determine the total current.
Combination circuits utilize both series and parallel connections. To analyze them, we apply Ohm's law systematically to each part. While complex, a methodical approach using Ohm's law at every step allows us to fully understand the behavior of the circuit.

Calculating Electrical Power

Grasping the nuances of electrical power is a cornerstone of the skilled journeyman's expertise. The interplay between voltage, current, and resistance drives power calculations that inform critical system design decisions. Through detailed explanations, sample calculations, and practical examples, readers will gain the working knowledge of power principles needed to size equipment, conductors, and protective devices efficiently.

We begin at the basics, reviewing the standard DC circuit definitions that form the backbone of electrical calculations. Voltage (V) constitutes the electric potential difference that causes current flow, measured in volts. Current (I) is the movement of electric charge, measured in amperes. Resistance (R) opposes current flow, measured in ohms. Ohm's Law fundamentally links these variables, expressed as V = IR. This formula empowers technicians to determine any variable when the other two are known.

Building upon Ohm's Law, power (P) equals voltage multiplied by current (P=VI). Power represents the rate of electrical energy usage or work performed, measured in watts. Some key characteristics of electrical power include:

Real power performs desired work like heating, lighting, or motor motion

Reactive power involves stored energy cycling in inductors/capacitors

Apparent power is the vector sum of real and reactive power

Power factor describes efficiency as real power divided by apparent

Journeymen must have intimate familiarity with these concepts when designing and troubleshooting electrical systems.

The basics provide a strong foundation, but proficiency requires going deeper. In AC circuits, conductors exhibit both resistive and reactive properties. This reactive impedance results from inductance or capacitance, causing voltage to lag or lead current. Effective impedance Z depends on the vector sum of resistance and reactance.

In purely resistive circuits with a unity power factor, apparent power equals real power. But with reactive loads, apparent power exceeds real power due to circulating non-productive currents. As an example, consider a 60W light bulb powered by a 120VAC source. The resistive bulb would draw 0.5A, giving:

P = VI = 120V x 0.5A = 60W

S = P = 60VA

But inserting a 1mH inductor adds 10Ω of reactance. The reactive and resistive drops now vectorially sum to 100V. The current increases to 1A even though bulb power remains 60W.

P = VI = 120V x 1A x PF = 60W

S = V x I = 120V x 1A = 120VA

This demonstrates why both genuine and apparent power dictate conductor sizing. Control wiring based on real power alone risks under-sizing and overheating.

Power factor also impacts efficiency, with values nearer to 1 being ideal. Power correction capacitors can offset lagging reactive loads. Leading power factors caused by capacitive circuits likewise require correction. Journeymen must be adept at power factor analysis and improvement measures.

We now move from theory to practical calculations. Consider sizing a service for a small office containing:

(3) 1000W Electric heaters

(2) 400W Computers

(10) 60W Light Fixtures

(1) 1500W Laser printer

(1) 1200W Microwave

(1) 300W Ventilation fan

First sum the real power for each load:

(3 x 1000W) + (2 x 400W) + (10 x 60W) + 1500W + 1200W + 300W = 11,460W

Allowing for spare capacity, select a 15kVA transformer. The total load power is 11.46kW. As all loads are resistive, the power factor is unity. The apparent power equals real power:

S = P = 11,460VA

For 480V primary voltage, the full load secondary current is:

I = P/V = 11,460W / 240V = 47.8A

This allows correct sizing of the secondary conductors and service equipment. Similar load calculations inform the selection of utility services and transformers.

This example demonstrates the crucial role power computations play in design and installation. Additional concepts like kilowatt-hours for energy usage, surge power ratings, and 3-phase calculations build further competency.

Practical Applications of Basic Electrical Principles

To truly comprehend electrical systems, you must first understand the fundamental concepts of voltage, current, resistance, and power.

Voltage, measured in volts, is the potential energy that pushes electrons through a circuit. Imagine a waterfall - the higher the waterfall, the more potential energy water has. Now think of electrons as the water. Voltage is like the "height" of the waterfall, generating the force to move electrons.
Current, measured in amps, is the flow rate of electron movement. Using the waterfall analogy, current is like the volume of water flowing over the edge each second. A high current means lots of electrons are moving; a low current is a slower flow.
Resistance, measured in ohms, opposes the flow of electrons. Resistors absorb voltage and limit current. Going back to the waterfall, resistance is like large boulders in the water. More boulders slow the water flow. More resistance limits the electron flow.
Power, measured in watts, is the rate of electrical energy consumption. Using P=IV, power equals current multiplied by voltage. Think of a lightbulb - a 60W bulb uses 60 joules per second. More power means more energy used.
Ohm's Law links these concepts: V = IR. For example, a 10-ohm resistor in a 12V circuit will have a current of 12V / 10 ohms = 1.2A. You can also calculate resistance from voltage and current: R = V/I. The key is manipulating Ohm's Law to find any missing variable.
Kirchhoff's Laws allow for analyzing multi-loop circuits. Kirchhoff's Current Law states current going into a node equals the current leaving it. Kirchhoff's Voltage Law states the total voltage around a loop is zero. Always approach complex circuits one step at a time using these rules.
Kirchhoff's Laws are fundamental for analyzing multi-loop circuits. They allow you to methodically work through complex systems step-by-step.
First, a "node" refers to a connection point in a circuit where two or more components meet. For example, if wires from three different branches in a circuit are connected at one point, that junction is considered a node.
Kirchhoff's Current Law involves analyzing the current flowing into and out of nodes. It states that the sum of all currents entering a node must equal the sum of all currents leaving the node. In other words, current is conserved at a node - what flows in must flow out.
To utilize Kirchhoff's Current Law, designate a flow direction for the current in every conductor linked to a junction. Calculate the total of currents entering the junction, and then the total leaving it. Equate both totals to determine the value of the unknown current.
For example, consider a node with three wires connected where Wire 1 has 5A flowing in, Wire 2 has 3A flowing out, and Wire 3 has an unknown current. Using Kirchhoff's Current Law:
Current into node = Current out of node
5A (Wire 1) = 3A (Wire 2) + I (Wire 3)
So the unknown current in Wire 3 can be found as 2A.
Kirchhoff's Voltage Law states that the total voltage around any closed loop in a circuit must equal zero. In other words, the voltage drops and rises around a loop must balance out.
To apply this law, tour the loop and sum the voltage differences across each component. If any voltage is unknown, you can set the overall sum equal to zero and solve for the missing value.

Review and Self-Assessment Questions

Reviewing key concepts and testing your knowledge is crucial for exam success.
Thoughtfully work through the following examples. If you grasp the concepts quickly, you are on the right track. For any missed or confusing questions, reread the relevant sections until the material clicks. Don't lose heart if some questions seem tricky at first - with diligent review, they will start to make sense. The goal is to pinpoint and shore up areas needing improvement.

Consider this scenario:

A series circuit has a voltage source of 120 V and three resistors with the following resistances: 10 ohms, 15 ohms, 20 ohms. What is the overall resistance of the circuit?
To find total resistance in a series circuit, simply sum the individual resistances:
Rtotal = 10 ohms + 15 ohms + 20 ohms = 45 ohms
Now calculate the current in the circuit using Ohm's Law. The voltage is 120 V and the total resistance is 45 ohms:
I = V/R
I = 120 V / 45 ohms
I = 2.7 A

Next question:

An electric heater requires 220 V to operate and draws 15 A of current. What is the power consumption of the heater?
Use the power equation:
Power (Watts) = Voltage (Volts) x Current (Amps)
Power = 220 V x 15 A
Power = 3300 W

Moving on:

A circuit has three branches with the following currents: Branch 1 has 5A flowing through, Branch 2 has 2A, and Branch 3 has 9A. How do you calculate the aggregate current emanating from the power supply and entering the junction that unites the three circuits?
This requires applying Kirchhoff's Current Law. The total current into a node must equal the total current flowing out.
Current in = Current Branch 1 + Current Branch 2 + Current Branch 3
= 5A + 2A + 9A
= 16 A

Keep challenging yourself with questions that test conceptual knowledge and practical calculations. The more experience analyzing circuits, troubleshooting issues, and solving for unknowns, the more automatic your electrical fluency will become.

CHAPTER 2

ELECTRICAL CIRCUITS

Understanding Series Circuits

Connecting components in a series circuit is common in electrical engineering and requires a solid understanding of how current and voltage behave in these configurations. This foundational knowledge will prove invaluable throughout your career as an electrician.

When components are wired in series, the current passes through each one sequentially. The same current flows through each resistor or other device. However, the voltage gets divided across each element. The total voltage supplied by the source gets progressively "dropped" in portions according to the resistance value of each component.

Consider a circuit with a 120V supply and three resistors in series: a 10Ω, 50Ω, and 100Ω. The current throughout is fixed at 12A based on the total circuit resistance and Ohm's Law. But the voltages divide like this:

10Ω resistor: V = IR = (12A)(10Ω) = 120V

50Ω resistor: V = IR = (12A)(50Ω) = 600V

100Ω resistor: V = IR = (12A)(100Ω) = 1200V

The voltages add up to a total of 120V. This potential divide effect is highly significant to understanding series circuits.

When analyzing series circuits, you can also apply Kirchhoff's Voltage Law. This states the sum of all voltage drops around a closed loop must equal the total voltage supplied. So in this example:

120V (Source) = 120V (10Ω resistor drop) + 600V (50Ω drop) + 1200V (100Ω drop)

Practice calculating the total resistance for components in series by adding their resistances. Get comfortable determining current with Ohm's Law using the total circuit resistance. Know how to divide the voltages. These skills will prove invaluable for series circuit analysis.

You may also encounter complex networks containing multiple series circuits within parallel branches. Methodically approach each series loop individually using KVL (Kirchhoff's Voltage Law) and resistance addition. Then relate them to the whole using current division rules at each node. Take the time to draw clear circuit diagrams and work through them one step at a time.

With dedication, series circuits and the interplay of current, voltage, and resistance in these configurations will become second nature. Gaining mastery over fundamental concepts like series analysis is the key to success as you embark on your electrical engineering career. The hands-on work of an electrician relies on being able to fluently apply the theory.

Exploring Parallel Circuits

Parallel circuits provide multiple paths for current to flow through simultaneously. While the voltage is constant across parallel branches, the current divides based on the resistances of each path.

To analyze parallel networks, apply these pivotal principles:

Voltage is equal across all parallel branches. Each component sees the full voltage.

Current divides between parallel branches depending on their resistances. More current flows through paths of lower resistance.

Total current is the sum of the branch currents (Kirchhoff's Current Law).

Total resistance is discovered by computations with reciprocal resistances.

Consider a circuit with three parallel resistors:

- R1 = 10 Ω
- R2 = 50 Ω
- R3 = 100 Ω

With a 120 V source, the voltage across each resistor is 120 V. Using Ohm's Law, the branch currents are:

- I1 (through 10 Ω resistor) = V/R = 120/10 = 12 A
- I2 (through 50 Ω resistor) = V/R = 120/50 = 2.4 A
- I3 (through 100 Ω resistor) = V/R = 120/100 = 1.2 A

The total current is the sum of the branch currents:

Itotal = I1 + I2 + I3 = 12 A + 2.4 A + 1.2 A = 15.6 A

To find the total parallel resistance, use: 1/Rtotal = 1/R1 + 1/R2 + 1/R3

This gives 1/Rtotal = 1/10 + 1/50 + 1/100 = 0.0642 Ω^-1
So Rtotal = 15.6 Ω
As another example, consider a parallel circuit with:

- R1 = 560 Ω
- R2 = 390 Ω
- R3 = 820 Ω

With a 210 V source:

- I1 = V/R = 210/560 = 0.375 A
- I2 = V/R = 210/390 = 0.538 A
- I3 = V/R = 210/820 = 0.256 A

Itotal = 0.375 A + 0.538 A + 0.256 A = 1.169 A
For the total resistance:

- 1/R1 = 1/560 Ω = 0.00179 Ω^-1
- 1/R2 = 1/390 Ω = 0.00256 Ω^-1
- 1/R3 = 1/820 Ω = 0.00122 Ω^-1

1/Rtotal = 0.00179 + 0.00256 + 0.00122 = 0.00557 Ω^-1
So Rtotal = 180 Ω
Methodically applying parallel analysis principles to practice problems is key for building expertise. Take the time to diagram and work through methodically.

Combination Circuits and Their Applications

Analyzing combination circuits containing both series and parallel elements is essential for real-world electrical engineering. Approaching these hybrid networks strategically will build key skills for complex system design and troubleshooting.
Any circuit containing both series and parallel branches is a combination circuit. Although initially intricate, combination circuits can be decoded by breaking them down into simpler series and parallel configuration. Reduce larger schematics down into individual series circuits and parallel nodes.
For example, consider a circuit with two voltage sources and five resistors:

- R1 = 50Ω and R2 = 75Ω in series
- R3 = 100Ω and R4 = 150Ω in parallel
- R5 = 125Ω in series with the R3/R4 combination

First, calculate the total series resistance of R1 + R2 = 50Ω + 75Ω = 125Ω.
Then find the parallel resistance of R3 and R4 using reciprocal resistance math:
1/100 + 1/150 = 0.01 + 0.0067 = 0.0167
So 100Ω || 150Ω = 60Ω
Now this 60Ω result is in series with R5 = 125Ω.
So 125Ω + 60Ω = 185Ω
By methodically combining the series and parallel pieces, we simplified this combination circuit down to two key resistances. The same approach applies to more elaborate networks.
Combination analysis also relies heavily on Kirchhoff's Laws. Use Kirchhoff's Current Law (KCL) to analyze the currents at each node. And apply Kirchhoff's Voltage Law (KVL) to determine voltage drops around multiple loops.
For the example circuit with:

- R1 = 50Ω and R2 = 75Ω in series
- R3 = 100Ω and R4 = 150Ω in parallel
- R5 = 125Ω in series with the R3/R4 parallel combination

First, simplify the parallel and series groupings:
R1 + R2 = 50Ω + 75Ω = 125Ω
1/R3 + 1/R4 = 1/100 + 1/150 = 0.0167 → R3 || R4 = 60Ω
So R5 (125Ω) is in series with the 60Ω result.
Therefore, the total equivalent resistance is 125Ω + 60Ω = 185Ω
Now apply Kirchhoff's Laws:
Using Kirchhoff's Current Law (KCL) at the node where R3 and R4 join:
Current through R3: I3 = V/R3 = V/100Ω

Current through R4: I4 = V/R4 = V/150Ω
I3 = I4 (due to parallel configuration)
So if voltage V = 120V:
I3 = I4 = 120V/100Ω = 1.2A
KCL: I3 entering = I4 leaving = 1.2A
Therefore, KCL is satisfied.
Using Kirchhoff's Voltage Law (KVL) around the full loop:

- Vs is the source voltage
- V1 is the drop across R1
- V2 is the drop across R2
- V3 is the drop across R5
- V4 is the drop across the R3/R4 parallel combination

KVL states:
Vs - V1 - V2 - V3 - V4 = 0
If Vs = 120V, then:
120V - (I1R1) - (I2R2) - (I3R3) - (I4R4) = 0
120V - (50Ω)(I1) - (75Ω)(I2) - (125Ω)(I3) - (60Ω)(I4) = 0
Therefore, KVL is also satisfied with this combination circuit example.
Thoughtfully working through practice circuits is crucial. Visualize and diagram each portion. Check your understanding by building example circuits with test voltages and resistor values.

The Role of Resistors and Their Impact

Resistors are foundational components that limit current flow in circuits. Though deceptively simple, their applications span from current control and voltage division to sensing, signal conditioning, and precision measurements. Developing resistor intuition will prove invaluable.
Resistors oppose current according to Ohm's Law: R = V/I. For a given voltage, higher resistance values limit current, while lower values allow more current to flow. Consider two resistors in a 120V circuit:
10Ω resistor: I = V/R = 120V/10Ω = 12A
1kΩ resistor: I = V/R = 120V/1000Ω = 0.12A
The 1kΩ resistor allows 100 times less current, illustrating the current-limiting effects of high resistances.
Resistors also enable controlled voltage division. Placing resistors in series divides an input voltage into smaller outputs. A simple example is a voltage divider with two resistors:
Input voltage = 10V
R1 = 1kΩ
R2 = 500Ω
The voltage divides based on the ratio of R2/(R1+R2).
Here, R2 = 500Ω, R1 + R2 = 1500Ω.
So the output voltage = 10V * (500/1500) = 3.33V.
This voltage division effect is applied extensively, from biasing transistors to scaling signals.
Combining series and parallel resistors provides additional capabilities. Parallel resistors divide current among multiple paths, limiting total current. Complex resistor networks can control both voltage and current for sensitive components.
Resistors dissipate power as heat, based on the power equation P=I2R. High-wattage resistors are utilized for absorbing energy in power supplies, motor controls, and high-current applications.
Experience reveals appropriate resistor selection optimizes cost, performance, and reliability. Consider power handling, tolerance, temperature effects, and noise. Common specifications include:
Resistance value - limits current/divides voltage
Power rating - maximum power handling
Tolerance - precision of resistance value
Temperature coefficient - change over temperature.

Circuit Analysis Techniques with Practice Problems

Developing expertise in circuit analysis requires building a strong foundation in the underlying theory coupled with cultivating practical problem-solving skills. Moving beyond basic circuits requires a methodical approach and dedicated practice.

The foundation of circuit analysis is built on essential concepts that articulate the interconnections between current, voltage, and power.. Ohm's Law connects voltage and current in resistors. Kirchhoff's Laws examine complex networks - the Current Law (KCL) analyzes node currents and the Voltage Law (KVL) analyzes closed loops. The Power Law relates power dissipation to current and resistance.

Understanding these basic relationships allows for tackling more elaborate circuits. Consider this series-parallel combination:

- R1 = 10 Ω, R2 = 20 Ω, R3 = 30 Ω in series
- R4 = 15 Ω, R5 = 25 Ω in parallel
- Voltage source = 60 V

A systematic problem-solving approach is:

- Simplify series and parallel combinations
- Identify target values to calculate
- Apply Ohm's Law and Kirchhoff's Laws methodically

Check work thoroughly before moving to the next stage

For this example, first calculate the total series resistance:

R1 + R2 + R3 = 10 Ω + 20 Ω + 30 Ω = 60 Ω

Next, determine the parallel resistance:

- 1/R4 + 1/R5 = 1/15 Ω + 1/25 Ω = 0.067 → R4 | | R5 = 15 Ω
- The two key resistances are now 60 Ω and 15 Ω.
- Apply Ohm's Law to find the total current:
- I = V/R = 60 V / (60 Ω + 15 Ω) = 0.75 A

After solving for the equivalent resistances and total current, it is significant to check the validity by applying KCL and KVL. This verifies that the solution satisfies the key circuit laws.

For the example:

- R1 = 10 Ω, R2 = 20 Ω, R3 = 30 Ω in series, equivalent to 60 Ω
- R4 = 15 Ω, R5 = 25 Ω in parallel, equivalent to 15 Ω
- Total resistance = 60 Ω + 15 Ω = 75 Ω
- Total current = 0.75 A

To check with KCL, examine the currents at each node:

At the node joining R1, R2, R3:

- Current through R1: I1 = V/R1 = 60 V/10 Ω = 6 A
- Current through R2: I2 = V/R2 = 60 V/20 Ω = 3 A
- Current through R3: I3 = V/R3 = 60 V/30 Ω = 2 A
- I1 = I2 = I3 = Total current of 0.75 A

KCL satisfied.

At the node joining R4, R5:

- Current through R4: I4 = V/R4 = 60 V/15 Ω = 4 A
- Current through R5: I5 = V/R5 = 60 V/25 Ω = 2.4 A
- I4 = I5 = 0.75 A

KCL is also satisfied with this node.

To check with KVL, analyze the voltage drops around the loop:

- Voltage source = 60 V
- Drop across R1 = I1*R1 = (0.75 A)(10 Ω) = 7.5 V
- Drop across R2 = I2*R2 = (0.75 A)(20 Ω) = 15 V
- Drop across R3 = I3*R3 = (0.75 A)(30 Ω) = 22.5 V
- Drop across (R4 | | R5) = (0.75 A)(15 Ω) = 11.25 V

- Total drops = 7.5 V + 15 V + 22.5 V + 11.25 V = 60 V

KVL is also satisfied around the loop.

Using KCL and KVL validates the systematically solved solution. This builds confidence and deepens understanding of circuit analysis techniques.

CHAPTER 3

WIRING TECHNIQUES AND CONDUCTOR SIZING

Standard Wiring Practices

Proper wiring forms the backbone of any electrical system. Following accepted standards ensures installations are safe, reliable, and robust over the long term. While specific requirements vary, some key principles apply universally. Developing sound wiring skills early will prove invaluable throughout your career.

One critical task is selecting the proper wire size, known as gauge. It is essential for conductors to be properly dimensioned to ensure they can handle the anticipated electrical loads and potential fault currents safely.. Undersized wiring overheats, while oversized waste materials. Refer to the National Electrical Code (NEC) ampacity tables to determine the proper wire gauge based on operating parameters.

Meticulously consider derating factors like ambient temperature, number of conductors bundled, and installation environment. For example, wires in conduits or buried underground run hotter and require thicker gauges than open-air installations. Leave margin for future upgrades and fault tolerance.

Choose wire insulation suitable for the operating voltage and environmental conditions. Common insulations include PVC, rubber, cross-linked polyethylene (XLPE), Tefzel, fiberglass, Kapton, and more. Evaluate temperature range, flexibility, abrasion resistance, moisture resistance, and cost tradeoffs. Use shielded cables in noisy environments to reduce electromagnetic interference. Apply sleeving on exposed conductors to avoid inadvertent shorts. Specify plenum-rated insulation in forced air handling spaces as required by code.

Neatly organizing and routing wires improves reliability, maintenance, and aesthetics. Use cable trays, conduits, raceways, sleeving, and strain relief to protect and guide conductors. Avoid placing wires near high vibration sources or heat sources over 60°C which degrades insulation over time. Prevent conductor abrasion against sharp sheet metal or machined parts using grommets, edge wrap, and adequate space. Follow minimum bend radius specifications to avoid cracking insulation and reducing current flow area over time.

Ensure all connections are mechanically secure, electrically continuous, and corrosion-resistant. Use solder, crimp, or screw terminals appropriately based on the application. Verify tightness to avoid high-resistance joints, applying strain relief as needed to reduce wire movement. Keep high-current power conductors physically separated from low-level signals to avoid interference. Confirm proper polarity across all conductors and terminations. Include some service loop slack to ease future maintenance or modifications.

Clearly label both ends of every conductor following a consistent scheme. Color coding often designates specific wire functions. Unique numbering distinguishes individual conductors in complex wire bundles. Durable labels placed within 6 inches of terminations can withstand wear.

Proper system grounding and bonding protect equipment and personnel. Separate ground conductors sink current surges, EMI, and static discharges safely. Bond all conductive equipment enclosures directly to the ground which equalizes voltage potential. Follow NEC guidelines for ground conductor sizing, routing, connections, and quantity.

Use wiring diagrams to plan runs and facilitate future troubleshooting. Verify all circuits systematically post-installation for continuity, isolation between conductors, and proper end-to-end operation. Over time, check and re-torque critical connections as wires settle. Adhering to sound practices ensures the system performs reliably for years to come.

Techniques for Accurate Load Calculation

Calculating electrical loads accurately is critical for properly sizing wiring, protection devices, and power sources. While estimating provides a starting point, precise load analysis requires a thorough accounting of all demand in a system. Undersizing can lead to nuisance trips or inadequate capacity during peak demand. Oversizing wastes capital on excess components and capability.

The foundation of any load study is a detailed inventory of equipment to be installed. Catalog specifications provide per-unit demand profiles under standard test conditions. Adjust values for unusual operating requirements or derating factors. Include generous allowances for unforeseen future additions.

For motors, the starting type significantly impacts peak demand. Inrush on across-the-line induction motors can spike up to 6 times the full load current. Soft start or VFD drives reduce starting surge. Take into account the duty cycle and the highest surrounding temperature, which can lead to an increase in winding resistance and the electrical current required.

Lighting systems draw steady resistive loads. Specify ballasts and drivers at 120% of the total connected load for a safety margin. Install occupancy and daylight sensors to reduce lighting energy use. Choose LED lights for efficiency and long life. Computers, electronics, and HVAC comprise an increasing percentage of commercial building demand. Tracking nameplate ratings assumes ideal efficiency - measure actual load with logging meters where possible. Derate transformer and UPS VA capacity 125% as a rule of thumb.

Evaluate duty cycle and load factors to determine average versus peak demand. Continuous loads like machinery can be summed directly. Intermittent hand tool use averages 50% time, so take 50% of the nameplate rating. Infrequently used appliances or seasonal loads apply lower factors.

Capture coincident loads that run concurrently. The fire pump may only need sporadic testing but draws substantial power when activated. Add simultaneous demands like elevators, security, and emergency egress. Statistical diversity applies to large numbers of similar cyclic loads.

Once the preliminary analysis is complete, validate and refine the findings. Conduct thermal imaging scans to identify hidden loads and issues. Take periodic current measurements at each panelboard and major load during typical operating conditions. Extrapolate peak demand from the data.

Poll end users to identify expected power-hungry additions or operational changes. Have vendors provide detailed startup/runtime profiles for large equipment. Reconcile preliminary estimates with real-world measurements. The most accurate load studies combine bottom-up inventory with top-down monitoring.

Safety factors account for inherent uncertainties in projections. Apply 15-25% margin on estimated peak demand asinsurance. A higher safety factor provides contingency at the cost of oversizing. Be conservative with mission-critical or electrically intensive facilities.

Document assumptions, calculations, and data sources thoroughly. Update the load study periodically and flag major occupancy or equipment changes. Load calculations require diligence but ensure electrical systems remain robust as demands evolve.

In summary, accurate load analysis requires:

- Detailed equipment inventory with specs
- Adjustments for unusual duty cycles
- Measurements to confirm nameplate demand
- Operational data to determine peak coincident demand
- Polling end users on expected additions
- Applying appropriate safety factors
- Clear documentation for future reference

Precise load calculations maximize system efficiency and safety. Consult relevant codes and facility operations to validate results. Invest time upfront to reap benefits in cost savings and reliability over the lifetime of the electrical distribution system.

Guidelines for Sizing Conductors

Selecting the proper wire size is a key step in any electrical design. Conductors must safely carry expected load currents without overheating or causing excessive voltage drop. Oversizing waste materials while under sizing risks equipment damage or fire. Consider the full operating conditions and follow best practices when choosing wire gauges.

The ampacity or current rating depends primarily on conductor size and insulation type. Thicker wires have less resistance and can sustain more current flow. Frequent sizes range from 16 AWG (smallest) up to 1000 kcmil or larger. The National Electrical Code (NEC) provides allowable ampacity in amperity tables for copper and aluminum wires with various insulations.

Evaluate the temperature rating of the wire insulation being put to use. Thermoplastics like THHN and XHHW are rated for wet locations up to 90°C. Use conductors with higher temperature ratings where possible to maximize ampacity. Size for the lowest rated component in the system.

Apply derating factors to account for real-world conditions. Bundled wires cannot dissipate heat as readily, reducing effective ampacity by up to 20%. Ambient temperatures above 30°C also decrease capacity. Consider voltage drop over long conduit runs. Allow margin for future loads and safety.

For branch circuits, size conductors for 125% of the continuous load plus 100% of non-continuous loads. The extra capacity handles temporary spikes. Use 80% of the circuit breaker amp rating as a maximum. For example, a 20A branch circuit would use #12 AWG good for 30A.

Feeders and mains may supply multiple circuits. Determine total estimated demand then add safety factors. Critical systems need larger contingencies - data centers often oversize by 50% or more. Balance cost versus risk when sizing.
The permissible length of a conductor is constrained by the voltage drop, which is dependent on the current of the load. Limit branch circuits drop to 3% max, feeders to 2%, and critical buses under 1% drop to avoid equipment issues or lighting flicker. Use the next larger standard size or specify higher voltage runs to reduce losses.
Short circuit ratings define maximum fault current capacity. Specify breakers and fuses that do not exceed wire SCCR. Use the utility fault study model to verify large distribution conductors. Protect sensitive equipment with current limiting fuses or reactors.
For motors, size for 125% of full load current for standard types. Larger motors have high starting surges - review NEMA tables. Specify thicker wires for long runs to limit voltage sag. Use dual ratings on VFD-driven motors. Consider NEC tap rules when sizing off motor branches.
Low voltage controls use 18-12 AWG typically. Critical data wire pairs may need shielding. Route away from EMI sources and use surge protection. Limit 24VAC transformer secondary loads to 80% of the rating.
Outdoors, increase wire gauge to account for temperature extremes. Use wet-rated jacketed cables. Bury underground runs at proper depth - rating decreases in conduit. Avoid aluminum wire unless thoroughly antioxidant treated.

Safety Considerations in Wiring

Electrical hazards pose serious risks to personnel and equipment. Careful planning and adherence to safety protocols help mitigate these dangers during installation and maintenance. Awareness of potential issues coupled with sound wiring practices safeguards both workers and the system.
Heat is a primary concern. Overcurrent from undersized wires or loose connections can rapidly reach dangerously high temperatures. Insulation damage, combustible materials, and prolonged contact with hot surfaces are all compound risks. Routine thermographic inspections of terminations, splices, and conductors quickly identify problems before they escalate. Properly sized wires, lugs, breakers, and disconnects limit temperatures during expected operation.
Arc flash when short circuits vaporize conductors also releases intense heat energy. Even momentary arcs can cause severe burns and ignition of nearby materials. Conduct an analysis of arc flash hazards for systems with medium and high voltage to measure the incident energy levels. Specifying current limiting fuses helps reduce prospective fault levels. Apply warning labels and establish proper PPE requirements based on hazard levels.
Electric shock remains an ever-present danger. Faulty wiring, damaged insulation, and improper grounding expose personnel to energized conductors. Ground faults on overloaded neutral conductors are insidious since breakers may not trip. Implement a solid equipment grounding system and test with hipots during commissioning. Instruct workers to use GFCI protection when possible and avoid working alone on live circuits.
Stored electrical energy can inflict injuries even after disconnecting power. Large capacitors take time to discharge - verify with a meter before touching terminals. Accessing medium voltage components like transformers or motors requires strict procedures to ensure isolation and lockout. Assume systems are alive until proven otherwise.
Toxic gasses or reduced oxygen levels are hazards in confined spaces with electrical equipment. Faults can produce deadly arc byproducts. Sealed battery rooms require ventilation during charging. Fiberglass insulation and dielectric fluids also pose respiratory risks if handled improperly. Provide fresh air exchange and exhaust in electrical rooms.
Water contact with energized conductors or equipment leads to electrocution or short-circuit fire hazards. Outdoor wiring requires weather-tight conduits and seals. Avoid routing new wiring below plumbing lines. Shut down and inspect electrical components after flooding. Ground fault protection is critical in damp areas.
Combustible material accumulations create fire fuel sources. Keep electrical rooms and conduits free of dust buildup, debris, and storage. Specify plenum-rated cables in ceiling return air spaces. Follow code separation rules for wiring near flammable liquids or gasses. Thermal overloads left unchecked degrade insulation over time.
Heavy conductor gauges pose lifting and bending hazards. Use mechanical aids for pulling large cables. Wear gloves to avoid cuts from sharp edges. Make gradual bends in fat wires to avoid cracking insulation. Suspend vertical cable trays securely at both ends and on all floors.
In summary, key electrical safety practices include:

- Checking for heat damage and high-resistance connections
- Performing arc flash analysis to specify PPE
- Ensuring effective system grounding and GFCIs
- Strict isolation procedures before accessing energized components
- Providing ventilation and monitoring air quality

- Keeping wiring dry and sealed from water intrusion
- Avoiding accumulation of combustibles
- Using safe handling techniques for heavy cables

Vigilance and safety-focused design, installation, maintenance, and operation of electrical systems reduce risks exponentially. Formal job briefings that cover hazards and mitigation steps keep workers aware. Safety and performance go hand in hand.

Case Studies and Problem-Solving Exercises

Case Study 1: Overheating Wires

You are called to investigate a 120V lighting circuit that started tripping breakers after years of reliable service. Checking the wire terminations in the panel and junction boxes, you find several hot connections that are warm to the touch. Measurements show the circuit now draws 18A with all lights on compared to a 14A nameplate rating. The conductors are aged #14 AWG copper was installed when the building was constructed.
Questions:

What factors likely caused the overheating connections?

- Older wires more susceptible to work hardening and loosening over time
- Accumulated dust/oxidation increasing resistance
- Additional loads added over the years exceed the original rating

What steps would you take to remedy the problem?

- Clean and re-torque all connections
- Check wires for cracks or damage and replace them if needed
- Review loads and upgrade wiring size if exceeding the 14A rating

How could this have been prevented during the initial installation?

- Proper torque and maintenance of connections
- Built-in safety margin by using #12 AWG for 15A circuit
- Confirming the rated load did not exceed 80% of the breaker

Case Study 2: Voltage Drop

A 100 hp motor starter contactor at the end of a 150 ft feeder run blows each time it tries to start the load. The motor previously operated fine on a shorter run. The conductors are 3/0 THHN copper installed in the conduit above the ceiling. Line voltage at the motor drops to 180V when starting versus the 208V panel rating.
Questions:

What is the likely cause of the excessive voltage drop?

- 3/0 AWG too small for 100hp motor starting current over 150ft

What mitigations would you suggest to address the issue?

- Install lower resistance cables, potentially parallel runs
- Connect the motor to a closer panel with a shorter feeder length

How could voltage drop calculations during design have caught this beforehand?

- Check the starting current on the motor spec sheet and do voltage drop calcs
- Ensure feeder size limits drop to 2% or less

Case Study 3: Arc Flash Incident

An apprentice electrician incorrectly opens a 480V MCC bucket while the supply is still energized, causing an arc flash incident. The technician suffers minor burns to his hands and arms while the arc blast damages the surrounding switchgear. The facility has no prior arc flash analysis completed or warning labels installed.

Questions:

What procedural errors contributed to this incident?

- Failed to confirm isolation and lockout before accessing energized equipment
- No PPE worn appropriate for potential arc flash hazards

What steps should the company implement to improve safety?

- Perform arc flash study and label equipment with hazards
- Improve LOTO procedures and training

How can PPE and warning labels help reduce arc flash risks?

- Ensure proper PPE is specified and worn for the task
- Warnings indicate the hazard level and remind workers of the risks

Practice Exercises:

A 3-phase, 480V, 60A motor load is located 200 ft from the panel. Determine the maximum wire size to limit voltage drop to 5%.

- Voltage drop = (2RI*L)/1000
- For 5% max drop: (20.12560*200)/1000 = 3V drop
- 3/0 AWG (~0.125 ohms/1000ft) meets the requirements

An office admin area requires 15 dual receptacle branches for computers and task lighting. Estimate the branch circuit and feeder size needed assuming 200A service.

- 15215A = 450VA total demand
- Feeder ~ 35A (125% * 450VA)
- Branches - #12 AWG for 15A circuits

A pulsed welding load will draw 50kA for 30 cycles when firing. Calculate the IR drop across 50 ft of 2/0 cable with 0.125 ohms/1000 ft resistance.

- I = 50,000A
- R = 50ft * 0.125 ohms/1000ft = 0.00625 ohms
- IR = 50,000 * 0.00625 = 312.5V drop

Covering sample responses and solutions reinforces the learning process. The depth of analysis helps develop critical thinking abilities.

CHAPTER 4
OVERCURRENT PROTECTION

Fundamentals of Overcurrent Protection

Overcurrent protection is an essential feature in electrical systems, designed to protect conductors and equipment against currents that exceed safe thresholds. Correct sizing and a selection of protective devices are essential to provide quality power and prevent hazards.

Circuit breakers rely on a current sensing element that trips the mechanism and interrupts flow when overload conditions occur. Thermal-magnetic breakers use a bimetal strip and electromagnetic coil in series. The trip time decreases as the overload increases following an inverse time curve. Molded case breakers are compact and enclosure-mounted. Drawout styles allow removal without de-energizing.

Breakers are available in standard and high interrupting capacities. Standard devices up to 10kA short circuit rating are suitable for most branch and feeder circuits. High-capacity molded cases and insulated case breakers handle up to 200kA for large distribution systems. Select breakers rated for the available fault current.

Fuses incorporate a calibrated fusible link designed to melt and open on sustained overloads or short circuits. Time delay types allow temporary inrush up to 10x rating. Fast-acting fuses quickly clear severe overcurrents within one-half cycle. Current limiting fuses reduce prospective let-through energy. Fuses are single-use and must be replaced after opening.

Coordination between series overcurrent devices prevents unnecessary upstream tripping on downstream faults. This selectivity is achieved by choosing devices with appropriate time-current curves. The objective is to ensure that only the protective device closest to the fault is activated. Lack of coordination increases downtime and hazards during troubleshooting.

Transformer primary fuses limit damage and arc flash risks by isolating downstream faults faster than the secondary breaker can act. Special breaker-fuse combinations provide selective coordination up to 200kA. Current limiting fuses may also be paired with medium voltage breakers and reclosers.

Ground fault protection disconnects circuits with leakage currents exceeding 5-30mA through alternate paths. Ground faults indicate insulation breakdown or energized equipment frames. GFCI breakers and receptacles protect personnel at 125V. GFPE provides broad ground fault detection on feeders. Isolating ground faults quickly prevents potential electrocutions.

Residual current devices monitor imbalances in line and neutral conductors caused by leakage. Opening within 30mA, they provide redundant protection with GFCI and insulation monitoring equipment. RCDs are faster than GFPE systems for personnel safety. 4-pole devices also detect neutral-ground faults.

Arc fault circuit interrupters (AFCIs) mitigate fire risks from arcing in damaged wiring and cords. Both branch/feeder and outlet versions are available. AFCIs detect the unique voltage and current signals caused by arcs. Combination AFCIs add grounded neutral sensing. Siemens, Eaton, and SqD offer AFCI breakers and receptacles.

With the wide selection of overcurrent protective devices, each system can be tailored for safety, continuity, and coordination. Careful product specifications matched to design requirements provide cost-effective protection against damage and disturbances.

Types and Characteristics of Fuses

Fuses are overcurrent protection devices containing calibrated fusible links that melt when subjected to sustained overcurrents. Understanding the different fuse types, critical specifications, and operating principles facilitates selecting the optimal fuse for an application. This chapter explores common fuse varieties and selection factors.

Standard time delay fuses feature a zinc, copper, or silver element that melts when overheated by overloads. They allow harmless inrush currents up to 10x rating but blow within seconds for higher overloads. Sizes up to 600A are available with voltage ratings up to 600VAC. Time delays up to 120 seconds handle motor starting spikes.

Fast-acting fuses use thinner fuse links that quickly interrupt severe overloads above 200% of rating. Melting occurs in under 0.1 seconds considering response time inversely proportional to overcurrent magnitude. Fast-acting fuses offer maximum short circuit protection but cannot withstand repeated inrush events.

Time delay fuses limit arc flash energy in short-circuit conditions. Inductive loads like motors produce high arc voltage so fast-acting styles are widely preferred for guaranteed safety. Both cylindrical and blade cartridge styles are common. Indicator windows or blown fuse flags denote opened elements.

Current limiting fuses incorporate special materials that increase resistance during faults. This quenches the arc and limits peak let-through currents to less than 10x rating. Faster clearing reduces arcing damage. Limiters cannot be reused but avoid cascade failures. They also mitigate arc flash hazards.
Semiconductor fuses handle the fast transients and high inrush associated with solid-state devices like drives and power supplies. Very fast-acting elements clear overloads quicker than conventional fuses. High I2t ratings withstand pulsed currents. The most frequent ratings are up to 600VAC and 200A.
High voltage fuses operate on medium voltage power systems between 601V and 69kV. They are substantially increased with higher interrupt ratings. Oil-filled designs immerse the fuse in dielectric fluid for arc quenching. Expulsion types vent gasses outward to extinguish arcs. Placement is critical in loop systems.
Fuse selection factors include voltage, ampere rating, interrupting rating, and physical size constraints. Verify that specifications meet or exceed the electrical system requirements with a safety margin. Evaluate time-current curves for coordination with upstream and downstream overcurrent devices.
Ampere rating determines the continuous current a fuse can handle without opening. This rating must exceed the maximum load current. Standard sizes increment from 1A up to 6000A+. Select the next higher standard rating above the circuit current.
Voltage rating depends on system voltage. 125VAC fuses apply for lighting and receptacle branch circuits. 600VAC fuses cover 208-480V panels and feeders. Use up to 35kV fusible devices on high voltage switchgear.
Interrupting rating defines short circuit resistance - minimum is 10kA, but fuses up to 200kA are available. This must exceed the bolted fault current based on power system analysis. Lacking detailed fault data, select 200kA for critical circuits.
Accurately specifying and applying fuses provides selective coordination and overcurrent protection tailored to the electrical system requirements. Consider fuse characteristics along with potential arc flash hazards when choosing between fuses and circuit breakers.

Types and Characteristics of Circuit Breakers

Circuit breakers are automatic switches designed to protect an electrical circuit from damage by excess current. There are several different types of circuit breakers, each with unique operating principles and characteristics. To select the right circuit breaker, match its voltage, current, interrupting ratings, and trip settings with the system's needs.
Molded case breakers (MCBs) are compact devices enclosed in a molded insulating housing. They are used in distribution panels, switchboards, and motor controllers up to 600V and 200A. Bimetallic thermal elements provide overload protection while electromagnetic coils trip for short circuits. MCBs are designed for fast tripping on 10-50X overloads. Frame sizes vary from 15-250A.
Miniature circuit breakers (MCBs) are designed to fit on 35mm DIN rails for industrial control panels and switchgear. They are available up to 125A and feature fixed or adjustable magnetic trip settings. The trip indication is shown by a toggle position. Miniature circuit breakers offer integrated accessories like auxiliary contacts, shunt trips, and alarms.
Insulated case breakers (ICBs) enclose the components in an insulated housing like MCBs. However, they have higher interrupting capacities from 10-150kA for large switchboards above 400A. Drawout designs allow removal without de-energizing. ICB frames range from 400-4000A at up to 600VAC.
Air circuit breakers (ACBs) rely on arc quenching in open air. They are applied in switchgear assemblies above 400A, up to 38kV. Open contact gaps generate high dielectric strength to interrupt faults. ACBs may include gas blast or magnetic blowout arc control. Larger distribution classes suit utility substations.
Molded case switches (MCS) lack overcurrent tripping but can act as manually operated switches and disconnects. Switching mechanisms are identical to MCBs but trip units are omitted. MCCBs can also be converted into non-automatic MCS types. Ratings match MCB limits.
Solid-state breakers replace the thermal-magnetic trip device with electronic current sensing. This allows adjustable overload pickups from 2-10X rating and configurable time-current curves. Ground fault and arc fault protection modules can be integrated into systems. Both Eaton and Siemens provide solid-state trip units with capabilities up to 150 amperes.
Fuse-break combinations integrate a fusible link overload sensing element with an air-break switch. They combine positive fuse overload protection with the ability to manually open circuits up to 600V/600A. Selection includes dual UL98 and UL248 standards listing.
Circuit breaker specifications like voltage, interrupting rating, and available accessories determine applicable uses. Choosing the proper trip rating and adjustment range provides selective coordination. Thermal magnetic units offer a simple, proven method to guard against overloads and faults. More complex designs enhance precision and flexibility.

Application and Selection of Overcurrent Devices

Correctly applying and choosing the right overcurrent protective devices is critical for an electrical system's safety, reliability, and selective coordination. Overcurrent devices like fuses, circuit breakers, and relays must match the voltage and fault current levels clearly shown while providing optimized protection and life safety.

Voltage ratings on overcurrent devices should exceed the nominal system voltage with a safety margin. 480V panels fed from 480V 3-phase TN systems require 600VAC fuses and breakers. Select 125% of maximum voltage. For 480V switchgear, choose 800V/1000V devices. NEMA sizes breakers for use up to 121% of rating. Voltage ratings for fuses are more conservative at 200%.

The continuous ampere rating must exceed the current of protected equipment in all use conditions. Branch circuit breakers are typically 15-30A for receptacles and lighting. Large motors require fuses or breakers 1.25-1.4x full load amps. Size feeder protection to carry starting currents of all connected loads. Always round up to the next standard size.

Interrupt ratings indicate short circuit current withstand for fuses and breakers. Standard devices are usually 10kA or 22kA. For high fault availability >50kA, specify breakers >65kA interrupt rating and 200kA limiters. Lacking detailed fault data, choose 200kA for critical circuits and 42kA for general branch devices.

Selectivity involves cascading overcurrent device time-current curves to isolate faults while minimizing outages. Use fast-acting devices like current limiters on large feeders and standard dual-element types downstream. Zone interlocking, differential relays, and ground fault protection also enhance selectivity.

Fuses and molded case circuit breakers offer fixed, inverse time-current curves. Adjustable breakers allow setting pickup and delay to achieve coordination. Select trip units that align operating bands without gaps or overlaps. Graphical plotting confirms proper sequence.

NEC requirements govern aspects like branch circuit MOCP sizes, the maximum number of breakers per panel, and requirements for local disconnects on equipment like HVAC units and motors. Follow all relevant codes to ensure safety and regulatory compliance.

Location affects the environment and duty of overcurrent devices. Indoor cool, dry conditions allow lower enclosures. Outdoor, wet industrial settings may mandate NEMA 4X housing with external operability. Consider temperature derating above 30°C. Access to testing and resetting is also critical.

Installation practices impact performance. Use torque wrenches for lug connections to avoid loose wires and hot spots. Maintain conductor insulation integrity. Verify that terminals are suitable for the wire gauge and number. Ensure correct alignment of fuse blocks and provide adequate support in areas prone to vibration.

Overcurrent device selection balances protection, arc flash risk, selectivity, and reliability. Apply adjustable trip breakers for overload tuning capability. Fuse-breakers and fast-acting limiters are optimal for arc flash reduction. Follow industry standards and use time-current curves for coordination. Match specifications carefully to the electrical system. Proper application fulfills protection goals safely and economically.

Scenario-Based Examples and Solutions

Determining appropriate overcurrent devices for specific electrical systems is best illustrated through practical scenarios. Analyzing example situations develops the critical thinking skills needed to make informed protection decisions. This chapter walks through sample cases and suggests solutions to reinforce key learnings.

Scenario 1:

An industrial facility is upgrading from a 150 kVA transformer to a 300 kVA unit to support new production lines. The existing main breaker panel and motor control centers will remain in use. What should be considered regarding overcurrent device selection for the transformer primary and new loads?

- Confirm the transformer's primary and secondary ampacities and choose appropriate rated feeder protection
- Check if the new transformer kVA fault availability requires higher interrupting rated devices than presently installed
- Size the secondary main breaker for the transformer rating, consider 125% rule
- Evaluate if MCC buckets and branch circuits can handle the new motor starting currents
- Review the overall system coordination with any new devices using time-current curves

Scenario 2:

A hospital's existing electrical gear does not have arc flash warning labels. Maintenance staff report the equipment is originally from 1980. There have been minor upgrades but no new overcurrent study. What actions would you recommend to improve safety?

- Perform a new short circuit and coordination study based on current equipment
- Calculate the arc flash incident energy levels and compare them to PPE requirements
- Apply warning labels per NFPA 70E and update PPE standards as needed
- Consider retrofits like high-resistance grounding to limit arc flash hazards
- Provide arc flash safety training to electrical personnel

Scenario 3:

At an automotive plant, the 60A feeder breaker for a critical 35 hp pump motor trips regularly at start-up but has operated for years without issue. Voltage drop checks find the problem is transient starting voltage dips. What are possible ways to address this?

- Upgrade wire size to reduce starting voltage drop
- Replace breaker with adjustable trip delay to ride through starting
- Specify a breaker/fuse combination device for better inrush tolerance
- Install a variable frequency drive to soft-start the motor
- Connect the motor to a nearer panel with a shorter wire run

Practical examples require integrating theory with experience-based judgment. Discussing realistic scenarios builds analytical abilities while reinforcing core protection principles. Both simple and complex situations warrant examination to expand working knowledge. A broad scenario base keeps protection skills sharp.

CHAPTER 5
DISTRIBUTION EQUIPMENT

Introduction to Transformers and Their Function

Transformers are essential components in electrical transmission and distribution systems, allowing AC voltages can be effectively increased or decreased as required for the transmission of power. Their flexible voltage transformation capability relies on the principles of electromagnetic induction between isolated windings wrapped around a common core.

Operating Principles

The basic operating principle of a transformer utilizes two electrically separate coils placed near to each other. The primary winding connects to the AC power source. When energized, it produces an alternating magnetic flux in the transformer's core. This changing magnetic field induces a corresponding EMF and current flow in the secondary winding due to the mutual inductance. Power is transmitted between the windings without an electrical connection.

- AC excitation of the primary winding creates a fluctuating magnetic field in the core
- Changing the magnetic field induces voltage and current in the secondary winding
- No direct electrical connection between windings provides isolation
- The turns ratio between windings determines voltage transformation

Key Design Factors

Several key design elements optimize transformer performance. The core guides and channels the magnetic flux using high-permeability materials that minimize losses. Common materials include:

- Laminated steel - low hysteresis and eddy current losses
- Ferrite
- Amorphous metal alloys

Windings must be thoroughly insulated from each other and the core. Typical insulating materials include:

- Polymer films
- Resin-impregnated paper
- Enamel coatings

Careful construction techniques and impregnation processes enhance the insulation system's dielectric withstand capabilities.

Transformer Types and Applications

Transformers are formed in various configurations tailored for different voltage transformation applications:

- Step-up - More secondary winding turns result in higher secondary voltage
- Step-down - Less secondary turns results in lower secondary voltage
- Multi-winding - Multiple isolated windings with flexible tap arrangements
- Specialty - Custom designs optimized for size, efficiency, harmonics suppression, etc.

Applications include:

- Transmission system - High power bulk transport of electricity
- Distribution grid - Parceling power to end users
- Within facilities - Tailored to machine or equipment needs

Voltage Transformation Equation

The relationship between primary and secondary voltages is defined by the turns ratio as shown in this formula:

$V_s / V_p = N_s / N_p$

Where:

V_s = Secondary Voltage

V_p = Primary Voltage

N_s = Number of Secondary Turns

Np = Number of Primary Turns

Understanding Voltage Regulators

Voltage regulators are essential devices that continuously adjust output voltage to loads to maintain a steady, consistent power flow, even when the input supply voltage fluctuates. This capability protects sensitive electronic equipment and prevents disruptive electrical anomalies.

Operating Principles

Voltage regulators maintain consistent load voltage by monitoring the input voltage, contrasting it with a stable reference, and modulating the output resistance accordingly. The basic operating principle is:
Vout = Vin * (R2/(R1+R2))
Where:

- Vout = Output voltage to load
- Vin = Fluctuating input supply voltage
- R1 = Variable internal resistance
- R2 = Load resistance

As Vin changes, R1 is frequently varied to maintain Vout at the target value. Fast reactive circuits allow precise real-time compensation within 1% or less.

Topology Options

Linear voltage regulators use active components like transistors in a variable resistive divider configuration to directly control output voltage. By doing this, you allow excellent response down to nanoseconds along with minimal ripple.
Switching regulators rapidly turn transistors on and off at high frequencies. By varying the pulse width, the average output voltage is controlled. Smaller filters are enabled. Efficiency and power density improve compared to linear types.
Some regulators utilize electromechanical components like motor-driven variacs and tap-changing transformers for coarser, slower control. Hybrid regulator designs also exist, combining electronic and electromechanical elements.

Essential Components

- Stable voltage reference - Provides the ideal baseline output voltage target
- Error amplifier - Compares actual output to the reference
- Pass element - Varies resistance to control current flow to the load
- Feedback circuit - Helps stabilize the control loop against oscillations

Key Performance Factors

Important parameters that determine if a voltage regulator suits an application include:

- Speed of response to disturbances
- Input and output voltage ranges
- Load current capacity and ripple limits
- Overall efficiency and reliability
- Overload and over temperature protections

By selecting regulators matched to the electrical system characteristics and load requirements, robust and stable power delivery can be maintained.

Typical Applications

Voltage regulators are heavily utilized in applications where consistent power with little fluctuation is critical, such as:

- Digital control and instrumentation systems
- Precision manufacturing equipment
- Medical diagnostic tools and instruments
- Telecommunications infrastructure

Proper voltage regulation is an essential factor enabling reliable operation of electronics-intensive systems and sensitive loads.

Capacitors in Electrical Distribution

Capacitors are crucial components in power distribution systems, providing vital benefits that keep electricity flowing smoothly to homes and businesses. Though small, these devices have an outsized impact on conditioning and stabilizing power delivery across expansive grids.

Prevalent throughout distribution networks, capacitors act like tiny reservoirs of electrical charge. They mitigate issues like voltage drops from line resistance, release reactive power to offset inductive loads and filter harmonic distortions from nonlinear devices. The result is a more optimal power factor and voltage regulation through all phases of system operation. Capacitors absorb power spikes and fill in dips, keeping voltage within tighter bands. They also reduce losses and efficiently maximize capacity.

Capacitors are often installed in banks (at substations) to provide these benefits across local feeders. Additional targeted placement occurs further down the lines nearer to points with heavy loads prone to voltage variability. Switching of capacitors can either be done manually or through automation, contingent upon the monitored conditions and metrics of power quality.. Specifying the right capacitor size and performance factors prevents overcorrection or underutilization.

Size for expected load and X/R ratio:

QC = (IVL) / (VC)

- QC = Capacitor reactive power (var)
- IVL = Load current and voltage drop
- VC = Capacitor voltage

Common capacitor types utilized in distribution applications include film, electrolytic, and power electronics-based designs tailored for durability and reactive power needs. Self-healing capabilities, current ratings, voltage withstand capabilities and insulation must all be evaluated for electrical grids' demanding environment. Cost, lifecycle maintenance, and control integration are also considerations for lasting value. Properly incorporating capacitors reduces waste, unlocks capacity, and enhances service.

From the massive transformers and cables down to ubiquitous pole-mounted units, capacitors serve quietly but critically alongside the backbone infrastructure that brings electricity to the masses. Their flexibility provides a tool for utilities to hone power delivery, avoid disruptions, and accommodate growing demand through optimizing existing grids rather than costly system-wide overhauls. Capacitors exemplify small components making a substantial impact on large-scale performance and reliability. Their role only continues to increase as distribution networks evolve.

Setting Up and Maintaining Distribution Equipment

Installing and sustaining electrical distribution equipment in the field requires careful planning, appropriate resources, and rigorous processes to ensure efficient, reliable, and safe power delivery.

When deploying new distribution equipment like transformers, capacitors, switches, and protective devices, utilities must first assess electrical system needs and service area, to load growth projections to define appropriate specifications. Environmental and right-of-way factors must also be evaluated in depth. Facility design, conductor sizing, layout, and components are selected accordingly. Proper tools, safety protocols, crews, and field supervision facilitate professional installation meeting codes and standards.

Correct settings, testing, and inspection verify that new assets integrate accurately with the grid. Monitoring and controls connectivity enables remote oversight. Utilities apply identification tags and update maps/records for future reference. A well-executed deployment and energization process pave the way for smooth operation.

Consistent preventative maintenance keeps equipment functioning optimally over decades of service. This involves patrols to check for damage, overheating, or irregularities. Testing procedures assess operation within specifications. Components like bushings, insulators, and gauges are tested for wear. Technicians perform minor repairs, lubrication, cleaning, or adjustments to address issues proactively.

For major overhauls, field crews utilize specialized tools to dismantle and rebuild aging transformers, switches, and other apparatus. Parts replacement renews performance. Upgrades are implemented as necessary. Meticulous documentation provides insight into asset conditions.

Utilities schedule outages to safely de-energize lines when maintenance is required. Coordination ensures minimal service interruptions. Crews don proper personal protective equipment and follow lockout/tagout procedures before working on lines or apparatus. Safety is the top priority.

From installation to upgrades to routine servicing, diligent maintenance sustains distribution assets for a reliable electricity supply. Crews utilize expertise and operational discipline to care for the vital equipment energizing communities year after year.

Troubleshooting Common Issues in Distribution Systems

Keeping power flowing to customers requires utilities to quickly identify, diagnose, and respond to problems arising in distribution systems. A combination of monitoring data, customer reports, and field intelligence guides strategic troubleshooting when issues emerge. Rapid pinpointed response prevents disruptions.

Remote Monitoring and Alarm Response

- SCADA systems with distribution grid sensors
- Real-time meter data on voltage, current, power quality
- Threshold alarms on abnormal loading, fluctuations
- Analyze trends and collect additional diagnostic data
- Characterize issues before dispatching crews

Addressing Customer Power Quality Issues

- Install temporary monitoring equipment at the service entrance
- Record voltage over time to quantify sags, noise, harmonics
- Compared to standards (IEEE 519, IEC 61000)
- Inspect loads and connections influencing quality
- Mitigate source issues like large motor starts

Outage Response and Restoration

- Check fuses, recloser targets, and operations counter
- Patrol and switching to isolate the faulted section
- Repair downed lines or damaged underground cables
- Replace blown fuses, faulty equipment like transformers
- Update and improve protection coordination

Proactive Equipment Monitoring and Maintenance

- Infrared, oil tests identify developing issues in transformers, switches
- Compare dissolved gas, insulation tests to baseline
- Address deteriorating assets before failure
- Employ reliability-centered asset replacement strategies

Systematic diagnosis guides targeted corrections that resolve distribution problems at the source. Quick response coupled with infrastructure reinforcements hardens grid resilience.

CHAPTER 6
MOTORS AND MOTOR CIRCUITS

Basics of Electric Motors

Electric motors are integral components across countless industrial, commercial, and residential applications. They efficiently convert electrical energy into mechanical motion and torque to drive generators, pumps, actuators, compressors, and more.

Motors contain magnets surrounded by wound coils of copper wire. Faraday's Law of electromagnetic induction states that an alternating current within coils generates a rotating magnetic field. This field interacts with the magnets to turn a central rotor and output shaft. Controlled acceleration to operating speed is accomplished through differing amounts of torque at startup versus continuous operation.

The two primary electrical motor types are AC induction and brushed DC motors. Induction motors utilize the rotating magnetic field principle with a squirrel cage rotor. The motor's speed depends on the line frequency. DC motors use commutators and brushes to power the rotating windings. Speed is proportional to armature voltage.

AC Induction Motor Overview

- Simple, rugged design with squirrel cage rotor
- Speed is determined by the number of poles and power frequency
- Output mechanical power function of torque and speed:

$P = T \times \omega$

T = torque (Nm)

ω = rotational velocity (rad/s)

- Losses include winding, core, stray load, friction/windage
- High starting torque capability

DC Motor Characteristics

- Wound rotor with commutator continually energized
- Output proportional to armature current and field flux:

$T = Kt \times Ia \times \Phi$

Kt = torque constant

Ia = armature current

Φ = field flux

- Separately excited or shunt wound fields
- Lower starting torque but variable speed capability

Key Selection Factors

- Required torque, speed, and horsepower ratings
- Continuous and peak load demands
- The motor speed-torque curve must exceed the load curve
- Efficiency ratings at typical duty cycles
- Temperature rise determined by insulation class
- Environmental protection level with enclosure

Motor Sizing and Application

Specifying the optimal motor involves matching requirements to capabilities across these factors. Additional components like gearboxes, variable frequency drives, and controllers enable further customization.

Designing Motor Circuits

Implementing proper power and control circuits is vital for electric motors to safely deliver optimized speed, torque, and protection across applications.

A well-planned motor circuit starts with appropriate supply voltage selection. The voltage rating should match expected operating levels to prevent overheating damage and provide full torque output. Power feeds utilize sufficiently sized conductors based on maximum current draw. Breakers, contactors, and fuses ensure overload protection.

For three-phase induction motors, confirming proper phase sequence prevents reverse rotation. Phase loss protection detects single phasing. Matching circuit impedance characteristics prevents voltage unbalance between phases. Proper grounding gives a safe path for faults.

Motor starters enable controlled acceleration by initially limiting voltage on startup. Reduced torque prevents abrupt jolts. Once at speed, connections transition to provide full power. For reversing duty, mechanical interlocks coordinate contactor sequencing. Automatic disconnection during power interruptions prevents unexpected restarts.

Adjustable speed drives provide a controlled AC frequency supply to the motor. Varying output frequency and voltage enable continuous speed regulation. Harmonic filters mitigate power quality issues from drives. Multi-motor applications require coordinated drives.

For DC motors, commutators and brushes conduct current to the armature windings. Connections must handle frequent sparking and wear. Speed control uses voltage regulators on shunt or separately excited fields. Series motors require fixed field weakening.

Carefully designed connections, protection, and controls enable smoothly harnessing the wide-ranging capabilities of electric motors for productive duty cycles in diverse applications.

Control Systems for Motors

Electric motors rely on properly designed control systems to safely regulate speed, torque, acceleration, positioning, and protection. Selecting and integrating the appropriate sensing, logic, and actuation components enables stable, responsive, and efficient motor operation tailored to the application.

The starting point involves appropriately sizing and rating controls to match the connected load. Evaluating motor nameplate data and mechanical requirements informs capacity needs and duty cycle constraints. Control circuits provide a scaled voltage or adjustable frequency supply for smooth starting. Solid-state soft starters and variable frequency drives offer advantages over traditional starters.

Sensors provide feedback on motor operation. Encoders, tachometers, and current transducers indicate speed, position, and electrical current for closed-loop control. Temperature switches guard against overheating. Vibration sensors can detect emerging mechanical issues. The data feeds to logic systems executing control algorithms.

Programmable logic controllers (PLCs) serve as centralized controllers for motors. Ladder logic programming coordinates sequence control, protection interlocks, and closed-loop regulation. Settings are adjustable for custom responses. PLCs readily network with HMIs and data systems. For standalone applications, motor drives can perform integrated control functions.

Output signals actuate contactors, switches, valves, and other components to orchestrate motor starting, speed, protection, and ancillary equipment. Contactors start and stop the motor. Drives or regulators adjust speed. The coordinated system achieves the desired motor response and capabilities.

Understanding control options and strategies is key to harnessing electric motor performance. Properly designed systems balance functionality, protection, and efficiency based on the application. With the right controls, electric motors can reliably deliver optimized motion generation and automation.

Protective Measures for Motors and Their Circuits

Electric motors represent major assets that enable productive processes across facilities. Implementing prudent protective measures provides safeguarding against damage while enabling reliability.

Specifying sufficiently rated insulation, enclosures, and temperature rise limits provides intrinsic protection tailored to the operating environment. Restricting motor loads to nameplate current ratings prevents overload. During short overloads, thermostats and variable frequency drives can temporarily lower speed to limit heating. For lengthy overloads, sizing conductors and starters for service factor currents add a safety factor.

Protective devices isolate motors from damaging issues in the supply system. Fuses and circuit breakers sized per code requirements disconnect overloads. Differential protection detects phase imbalances from uneven loading or supply issues.

Voltage monitors guard against under/overvoltage conditions. Ground fault relays prevent arcing damage. Surge suppressors filter power quality disturbances.
Embedded temperature detectors enable winding monitoring. Preset overload relays directly respond to excessive current. Resistance temperature detectors (RTDs) give continuous stator temperature data for predictive trends. Thermal imaging of enclosures also identifies hot spots preemptively. Vibration sensors help detect bearing wear or alignment problems.
Preventive maintenance extends protection. Testing insulation integrity per IEEE 43 standards identifies deterioration before failure. Lubricating bearings prevent friction damage. Inspecting air gap clearance and balancer condition avoids rotor impacts. Keeping filters clean promotes proper cooling.
A multi-layered strategy combining prudent design, monitoring, and maintenance safeguards electric motors from issues for maximum protection. Protecting these workhorse assets improves safety and reliability.

Motor Maintenance and Troubleshooting Techniques

Proactive maintenance and swift troubleshooting of electric motors are pivotal to extending service life and preventing disruptive downtime.

Preventive Maintenance

- Regular cleaning of air intakes and filters
- Lubricating and replacing bearings per schedule
- Testing insulation resistance per IEEE standards
- Inspecting connections and bus bars for tightness
- Thermography to identify developing hot spots
- Vibration analysis to detect imbalance or wear
- Tracking current, voltage, and temperature baseline data

Troubleshooting Common Issues

When problems arise, understanding failure modes and fault patterns speeds effective troubleshooting. Overheating indicates possible overloading, bad bearings, or inadequate cooling. Lack of starting torque or stalling suggests contactor or starter problems. Slow operation points to low voltage or worn brushes. Excessive vibration may signal unbalance or misalignment. Strategic testing isolates the root cause.
Once diagnosed, common remedies include cleaning and replacing air filters, adjusting brushes, changing bad bearings, realigning couplings, replacing worn contacts, fixing loose connections, and overhauling cooling fans. More serious and dangerous faults like damaged windings or rotors require motor rewinding or replacement. Record keeping informs maintenance planning and parts stocking.
With routine preventive care and knowledge of fault patterns, many motor issues can be avoided or quickly corrected to maximize uptime. Keeping detailed logs also enables predictive maintenance and design upgrades that improve reliability. Proactive motor maintenance pays dividends for continual performance.

CHAPTER 7

LIGHTING CIRCUITS AND CONTROLS

Design Principles for Lighting Circuits

When designing lighting circuits, follow key principles to optimize the performance, efficiency, and longevity of the system. Proper circuit layout, component sizing, voltage regulation, and integration with controls provide quality illumination precisely where needed. Planning circuits around established practices simplifies installation, enhances safety, and allows flexibility for future changes.

The first fundamental consideration is appropriate lighting levels for each space's purpose. Recommended illuminance values for office areas, manufacturing facilities, hospitals, and specific tasks should guide luminaire selection and placement. Build in flexibility to increase light levels later if needed. Determine the lumen output required to meet light-level goals based on room sizes and mounting heights.

Choose luminaires suited to lighting goals, aesthetics, and physical constraints. Compare photometric data like lumens per watt, optical control, luminance, and glare ratings. Modern LED fixtures provide excellent lumen output, color quality, and directional control in an efficient, long-lasting package.

Use the Dialux program to model the space and selected luminaires to optimize placement and aims for uniform light distribution and visual comfort. Avoid heavily biased lighting ratios. Locate switches conveniently while following electrical code requirements.

Circuit lighting logically to balance phase loading. Run alternate luminaires on separate circuits. Provide redundant switching or alternate power sources for egress and hazardous locations. Calculate voltage drop based on circuit length and load to avoid problems.

Incorporate controls like occupancy sensors, photocells, and manual dimmers to enhance efficiency. Evaluate interactions with other building systems to prevent interference or compatibility issues. Emergency lighting requires dedicated backup circuits.

Plan for easy access to above-ceiling components. Follow all NEC requirements and local codes for wire gauges, junction boxes, wiring methods, and circuit overcurrent protection. Use standard materials and wiring methods to simplify installation and maintenance.

The final lighting plans should completely define the right luminaires, circuiting, switching, and controls to meet application needs, efficiency goals, and electrical codes. Following sound design principles makes bringing concepts to reality straightforward and effective.

Determining Required Lumens

Use the lumen method:

Identify recommended illuminance (footcandles) for each area per IES guidelines
Calculate the square footage of the workplane
Multiply by required lumens per square foot based on room surface reflectances
Add 10-20% extra margin above the base calculation
For example, a 25' x 15' office with 50fc target has:
375 sq ft workplane
X 45 lumens/sq ft (for medium room reflectance)
= 16,875 lumens required

Selecting Luminaires

Compare photometric test data for luminance, luminaire efficiency (lumens/watt), optical control, color rendering index
Use tools like AGi32 to model luminaire placement and aim for ideal uniformity
Specify added accessories like baffles, lenses, or louvers if needed

Circuit Layout

Follow NEC table 210.24 for maximum loads per circuit type
Alternate luminaires between separate circuits for balanced phase loading
Check for voltage drop at the farthest circuit point:

Voltage Drop = 2 * Circuit Length * Load / Conductor Area
Must not exceed 3% dropout

Component Ratings

Specify wiring gauge based on ampacity tables for 40C ambient temp derating
Use standard wire colors and polarity throughout the installation
Size overcurrent protection at no more than 125% of continuous load
Select switches, relay panels, and dimmers suitable to lighting control needs
Thorough calculations and electrical knowledge inform lighting circuit design to create effective, efficient, and code-compliant lighting installations.

Installation Procedures for Lighting Systems

Proper installation of lighting systems requires coordinated planning and execution across electrical rough-in, fixture mounting, circuit wiring, testing, and control integration. Following structured procedures at each stage ensures quality workmanship, code compliance, and designed performance. Careful handoffs between tasks reduce rework from overlooked details or deficiencies.

The initial electrical rough-in prepares the building infrastructure to accept luminaires. Coordinate placement of junction boxes with lighting plans. Provide adequate support structures and lifts for suspended or high-mounted fixtures. Install access panels to allow maintenance access above ceilings. Dedicate neutral and ground conductors as specified.

Review photometric test reports and execute field aiming adjustments during mounting. Account for furniture layouts and sightlines. Verify suspended heights and orientation. Use factory-supplied hardware and methods only. Apply specified torque settings on bolts. Leave adjustments accessible where needed. Seal housing penetrations properly per UL listing.

Adhering to stringent wiring protocols involves marking conductors, confirming the polarity throughout the circuit prior to sealing, securing connectors and wire nuts tightly, and organizing cables tidily. Follow approved wiring diagrams and schedules. Test for continuity, grounding, and insulation resistance before energizing.

Group control wires securely and validate the functionality of switches, sensors, and dimming circuits before ceiling closure. Check conductor routing for pinch points. Inspect all terminations that are properly made. Thorough testing at this stage prevents issues later.

Complete final equipment adjustments and measurements after ceilings and furnishings are installed. Recheck alignments and distributions for uniform lighting. Replace any defective lamps or damaged luminaires. Verify rated light levels via spot measurements and adjust fixtures as needed.

Fully validate the operation of control systems including occupancy, photocells, presets, and interfaces. Record control settings and commissioning parameters. Resolve any compatibility or functionality issues through tuning and troubleshooting before sign-off.

Generate as-built drawings with marked-up changes and key installation details. Compile equipment manuals, warranties, and maintenance notes for the owner's facility staff. Offer training overview on proper operation of lighting systems and controls.

Consistent adherence to quality processes and attention to detail gives confidence in a robust lighting system installation that fulfills the intended design and performance for many years.

Electrical Rough-In

Size raceways and junction boxes per NEC fill requirements - calculate conductor cubic inches for 40% max fill
Use 3/4" conduit minimum; 1" flexible metal conduit for luminaires
Place boxes within 1 foot of lighting layout coordinates
Install suspended supports on 4-foot centers for continuous runs
Provide UL-listed through-branch circuit raceways for some applications

Luminaire Mounting

Review photometric report for task illuminance goals and test distances
Mark aim points on floors to properly locate peak candlepower
Maintain a minimum 3:1 maximum to minimum ratio on walls as per IESNA
Use spirit level on the mounting plate to prevent sag or drift over time
Follow manufacturer torque specs - typically 15-20 ft-lbs for 1/4" hardware

Circuit Wiring

Use standard color coding throughout for normal and emergency circuits
Connect only one conductor per terminal; no wire nuts in fixtures
Check polarity before capping each circuit using a phase tester
Use insulated crimp connectors for solid wire splices with 180-degree bend relief
Make tight but non-crushing terminations and re-check torque annually

Controls Integration

Isolate line voltage, VDC, and VAC control wiring in junction boxes
Label each control conductor by number and terminate systematically
Check/set dip switches consistently between all control elements
Confirm sensor coverage through walk-through testing after installation
Following precise, proven procedures for each stage of installation ensures lighting systems meet design performance goals and provide quality, energy-efficient illumination over the system's lifetime.

Overview of Modern Lighting Controls

Lighting controls go beyond simple switching to provide intelligent automation, optimization, and interfaces. Advanced technologies allow granular light tuning, occupancy adaptation, daylight harvesting, centralized monitoring, and network integration. Specifying the right control systems and strategies matches lighting performance to user needs while significantly reducing energy consumption.
Occupancy sensors detect space usage and vacancy to automatically turn lights on or off accordingly. Passive infrared, ultrasonic, and dual-tech models accommodate different applications. Sensitivity, time delays, and zoning prevent nuisance activation or deactivation. Partial-on keeps minimum illumination when vacant.
Photocells measure ambient light to modulate electric lighting output. Open loop systems at the luminaire adjust based on the local condition. Closed loop systems with ceiling or desktop sensors factor in overall room brightness for superior daylight harvesting.
Personal tuning systems empower occupants to set their local light level and color temperature preferences. Dimmers, wallstations, and handheld remotes provide adjustment. Some systems can store personalized presets. It enhances both ease and efficiency.
Color tuning controls alter the light color appearance throughout the day to support circadian rhythms and visual tasks. Preset or dynamic color changes are implemented via LED drivers or dedicated tuning modules. This enhances workplace wellbeing.
Schedule-based controls tailor lighting needs to the time of day and building usage patterns over the week. Astronomical time clocks adjust for seasonal daylight changes. Programmable schedules maximize efficiency during unoccupied periods. Manual overrides allow temporary changes.
Centralized control systems network luminaires, sensors, and building automation via digital communication protocols. Lighting control panels integrate and coordinate components enterprise-wide. Advanced software enables zonal control, load shedding, emergency response, and analytics.
Some specialty applications have unique control needs. Theatrical lighting requires intricate dimming, cues, and effects. Cleanroom lighting must avoid flicker that might impact processes. Control redundancies increase the resilience of critical egress or hazardous location lighting.

Occupancy Sensors

- Passive Infrared: Detects temperature changes from body heat up to 30 feet away. Use for open office areas, conference rooms, and restrooms. Provide a 180° detection pattern for optimal coverage.
- Ultrasonic: Detect Doppler shifts in reflected sound waves within a 25-foot radius. Effective for cubicles, aisles, and hallways. Reduce sensitivity to avoid false triggering from ambient noises.
- Dual Tech: Combines PIR and ultrasonic to prevent missed activations in noisy areas. Ideal for classrooms, auditoriums, and merchandising areas. Separate the sensors for wider coverage.

Program all sensors with a 15-30 minute delay after the last detection before switching off. Incorporate a brief fade warning before off. Specify partial-on to 50% brightness when vacant to assist navigation.

Photocell Daylight Harvesting

- Open Loop: Photocells on luminaires adjust fixture output based on daylight reading. Simple, low cost. Can cause uneven illumination.
- Closed Loop: Ceiling or desktop sensors network to uniformly calibrate output across multiple fixtures. Maintains ideal lighting balance.

Position sensors properly to measure overall ambient light, not glare or shadows. Program a setpoint of 25-40 footcandles at task level before dimming occurs. Set footcandle deadband to avoid oscillation around the setpoint. Hardwire photocells for reliable performance.

Color Temperature Tuning

- Preset: Change color temp according to preset schedules based on circadian patterns and usage.
- Dynamic: Tune CCT continuously across 2700K-5000K based on metrics like outdoor light.

Allow personal tuning of +/- 300K from preset to balance circadian alignment with user needs. Limit maximum CCT to 4000K to avoid harsh glare.

Centralized Networked Control

- Digital communications over WiFi, BACnet, and LONWorks enable enterprise-wide control, load shedding, and emergency lighting coordination.
- Pre-program lighting "scenes" tailored to activities and usage patterns for one-touch control.

Cloud-based software enables remote control, usage analytics, and data integration with other systems.

Commissioning and Analysis

Thoroughly test and adjust control settings during installation to optimize coverage, energy savings, and visual comfort.
Continuously re-calibrate sensors and update schedules over a full season cycle to match actual occupancy patterns and daylight availability in the space.
Track energy usage, create heat maps of occupancy patterns, and sensor effectiveness. Identify opportunities for added controls or re-configuration.
Advanced lighting controls provide the right light at the right time and place. Proper selection, strategic programming, and ongoing commissioning realize maximum energy reduction while enhancing occupant productivity.

Energy Efficiency in Lighting Design

Lighting design has experienced a transformation driven by advanced technologies and energy efficiency goals. Thoughtful application of modern luminaires, controls, and best practices creates quality, sustainable lighting.
LED lighting has sparked a revolution in capabilities. The solid-state technology enables superior color quality, dimming, control integration, and lifespan compared to outdated sources. Precision optics tailor illumination. Leveraging LEDs under the right specifications unlocks energy savings through efficiency and smart operation.

LED Technology Benefits:

- Superior color quality, dimming, and control integration
- Precision optics to tailor illumination patterns
- Extended lifespan compared to traditional sources
- Energy efficiency enables smart operation

Design begins by carefully evaluating the lighting task and users' needs. Light levels, color temperature, color rendering, optics, and distribution should align with the application and activities. Layering ambient, task, and accent lighting provides flexibility for varied scenes and occupancy. Minimizing overlighting avoids energy waste.
Designers finely tune illumination with fixtures, optics, reflectors, and walls. Window placement balances daylight. Dimming and occupancy controls reduce energy when full output is unnecessary. Daylight harvesting sensors lower electric light when ample ambient light exists. Scheduling matches lighting to occupancy patterns.
Good design also considers aesthetics and human factors. Avoiding glare, flashes, and strong shadows enhances comfort. Circadian-friendly warmer hues support overnight settings. The location of controls enables intuitive user adjustments. Commissioning ensures proper installation and operation.

With forethought and creativity, lighting design can harmonize visual performance, comfort, well-being, and remarkable energy savings. The human experience and energy efficiency both benefit from quality lighting.

Troubleshooting and Maintenance of Lighting Systems

Lighting systems require occasional troubleshooting and maintenance to sustain visual performance. A proactive, strategic approach prevents issues and extends service life.

Preventive maintenance establishes lighting health baselines and averts premature failure. Periodic inspections verify the proper aim and security of fixtures. Electrical connections must remain tight. Clean lenses and reflectors maintain light levels. Room surfaces should be cleaned to maximize reflectance. Replace lamps per manufacturer guidelines before burnout.

Troubleshooting begins by noting specific symptoms and ruling out obvious causes like switched-off breakers or lamps simply needing replacement. Dim lighting could indicate voltage drop, aged lamps, or dirty components. Flickering may signal a bad lamp, loose connection, or failing ballast. The color shift can derive from old lamps or voltage imbalance.

Logical isolation of the likely issue area accelerates diagnosis. Problems across multiple fixtures indicate a power or control system source. Hot relays or flickering points to electrical arcing and connections needing replacement. Issues in only one fixture suggest a component problem. Strategic testing and inspection narrow down the root cause.

Once diagnosed, issues like loose wiring, worn lamp sockets, out-of-aim floodlights, or dirty lenses can be directly corrected. Failed components require replacement with equivalent parts maintaining safety and performance. Beyond piecemeal repairs, systematic issues may warrant lighting upgrades or retrofits.

With methodical troubleshooting and preventive care, lighting systems can deliver decades of quality performance. Proper maintenance protects the investment in light for productivity and comfort.

CHAPTER 8
APPLIANCE CIRCUITS

Electrical Needs for Household Appliances

Providing proper electrical supply to household appliances enables safe, optimal performance. Careful load analysis during home design or remodeling ensures sufficient capacity and appropriate wiring. Key factors include voltage, amperage draw, circuits, and receptacles.

Comparing nameplate ratings to standard voltages avoids mismatch issues. The voltage drop experienced during load conditions must be computed, taking into account the dimensions and length of the wire..

Understanding the amperage needs of each appliance prevents overloading. Microwaves and dishwashers may draw 15A. Refrigerators need 5-10A. Electric stoves and water heaters can require 40-50A. The electrical panel and circuits must supply adequate current for simultaneous loads.

Key Design Considerations:

- Compare nameplate appliance ratings to service voltage (120V, 240V)
- Calculate voltage drop based on wire sizing and run lengths
- Sum amp loads; avoid overloading circuits
- Provide dedicated circuits for major appliances
- Install receptacles matched to appliance plug types

Service and Circuit Sizing:

Dedicated appliance circuits serve high-demand devices. Gas stoves often need a 120V, 20A circuit. Electric stoves require 50A circuits. Electric clothes dryers need a minimum 30A circuit. These warrant separate breakers instead of 15-20A small appliance branch circuits.

Receptacle types must match appliance plugs. Dryers use large 3 or 4-prong receptacles. Ranges need a 50A stove outlet. Electronic devices use 5-15 or 5-20 receptacles. GFCI protection is required for kitchen and outdoor outlets. Matching plugs prevent overloads from improvised adapters.

Voltage Drop Calculation:

VD = 2 * K * I * L / A
Where:
VD = Voltage drop
K = Resistivity of wire
I = Current (amps)
L = Length of wire (feet)
A = Cross-sectional area of wire (circular mils)

Receptacle Selection:

Dryers - 30A, 3/4-prong receptacle
Ranges - 50A, stove-type receptacle
GFCI - Kitchen, bath, outdoor outlets
Electronics - 5-15R or 5-20R
Proper appliance circuits ensure safe operation.

Special Considerations for HVAC Systems

Designing and maintaining HVAC systems for optimal performance requires special considerations regarding building load analysis, equipment selection, distribution design, controls, and maintenance procedures. Careful attention to these factors ensures the system meets comfort, efficiency, reliability, and cost objectives.

Accurately estimating heating and cooling loads based on climate, building characteristics, usage patterns, and internal/solar gains provides proper equipment sizing and design. Oversized equipment cycles inefficiently while undersized systems struggle to meet setpoints.
Equipment selection balances performance, energy efficiency, and lifecycle cost. High-efficiency units save operating dollars over time despite higher initial price tags. Two-stage compressors offer better humidity control. Inverter-driven technology enables variable capacity.
Ductwork and piping design factors in velocity, noise, pressure drop, and zoning need for effective air and water distribution without excessive fan and pump energy. Insulation levels, layout, sealing, and materials optimize efficiency.
Sophisticated building automation optimizes system control while enabling scheduling, setback, and monitoring. Strategies like demand-controlled ventilation adjust operations to occupancy patterns. Well-tuned controls prevent simultaneous heating and cooling.
Ongoing maintenance like replacing filters, cleaning coils, testing refrigerant charge, and verifying airflow keeps equipment in peak operating condition for comfort and efficiency over decades of service life.
With careful implementation of these HVAC best practices, commercial buildings benefit from excellent interior environments while avoiding wasted energy and excessive operating costs.

Wiring and Safety for Water Heaters

Providing adequate electrical supply and following safety protocols are critical when installing and maintaining residential water heaters. Proper wiring prevents hazards and enables optimal heater performance.
Electrical service must match water heater specifications. Storage tank heaters typically require a 30 amp dedicated circuit. Tankless heaters may need up to 150 amps depending on flow rate and temperature rise. The circuit breaker size depends on the amperage rating. 4 gauge copper wire is recommended for 50 amp circuits, increasing to 1/0 gauge for 100 amps.
Grounding provides a safe path for stray electrical currents to prevent shocks. The ground wire connects the water heater's metal frame to the ground bar in the main electrical panel. Homes should also have GFCI outlets within 6 feet of the water heater.
Combustion water heaters require an adequate air supply for the safe venting of exhaust gasses. Gas models need openings sized to BTU input. Direct vent heaters have dedicated ducting. Standard models draw combustion air from the room so adequate volume must exist.
Flammable materials should be kept away from the water heater. A drain pan catches leaks and connecting hoses should be steel braided. Inspection ensures venting remains intact. Temperature and pressure relief valves need periodic testing. Turning down thermostats prevents scalding. Proactive maintenance enhances safety.
Taking the time to correctly address wiring, venting, leaks, and materials greatly reduces risks of electrical faults, fires, carbon monoxide, and other water heater hazards through years of dependable service.

NEC Regulations for Appliance Circuits

The National Electrical Code (NEC) provides important regulations for designing and installing residential appliance circuits to ensure safe, dependable performance. Understanding key requirements helps achieve code compliance.
The NEC mandates dedicated circuits for major appliances expected to draw over 50% of a standard 15 or 20-amp branch circuit rating. This prevents overloads that can trip breakers or present fire risks. Electric ranges, wall ovens, clothes dryers, and air conditioner compressors require dedicated circuits.
These appliance circuits must be sized according to expected amperage loads. For example, electric dryers need a 30-amp circuit. Ranges require 40 or 50 amps depending on voltage. Over-current protection is incorporated through appropriately rated circuit breakers or fuses.
The NEC also governs receptacle types for various appliances. Dryer receptacles are distinct from standard 5-15 or 5-20 outlets. Ranges need specific 50 amp connections. Installers must follow designated configurations. (The NEC references "standard 5-15 or 5-20 outlets" when describing receptacles for general-use branch circuits. 5-15 denotes a grounded single receptacle rated for 15 amps and 120 volts. It has two vertical slots and a round ground pin. 5-20 also indicates 120 volts but with a 20 amp rating and has one vertical slot turned horizontally to prevent a 15 amp plug from being inserted. Both are ubiquitous outlets found throughout homes.)
GFCI protection is mandated for receptacles in areas near sinks or water sources, like kitchen countertops. This is critical for safety with plug-in appliances. Arc-fault circuit interrupters protect dangerous arcs in branch circuits. (GFCI stands for ground fault circuit interrupter. This is a safety device that quickly cuts power if it detects an imbalance in current flow. This protects against electric shock by interrupting flow when current leaks through water or a person's body. GFCIs monitor

the difference between hot and neutral and trip if greater than 4-6 milliamps. The NEC requires GFCI protection on receptacles near sinks, tubs, and other wet areas where plug-in appliance shock risks exist. Modern codes expand GFCI requirements to kitchens, garages, and many other locations.)

Adhering to NEC requirements for properly rated, dedicated circuits with listed overcurrent devices and receptacles ensures safe, long-lasting performance of high-load residential appliances.

Common Problems and Solutions with Appliance Wiring

Faulty wiring and electrical connections are common causes behind malfunctioning appliances. Being aware of some frequent issues and best practice solutions enhances troubleshooting and safety.

Loose connections anywhere along the electrical flow – at breakers, receptacles, or terminals – interrupt energy transfer and can generate heat and fire hazards. Tightening all contacts to specified torque values ensures robust, low-resistance paths.

Corroded contacts or terminals also degrade connections. Cleaning and coating with antioxidant paste helps maintain continuity. Damaged outlets into which plugs are inserted must be substituted with new ones.

Overloaded circuits shared by multiple appliances can trip breakers, especially with temporary spikes in a current draw during motor starting. Dedicated circuits are ideal for large loads like dryers, ranges, and air conditioners as required by code.

Undersized wiring restricts current below-appliance demand, impairing performance. Matching wire gauge to breaker amperage and run length maintains voltage within the appliance operating range.

Insufficient grounding pathways create potential shock or electrocution risks. Robust ground wires properly bonded throughout the system direct stray currents safely.

Diagnosing and rectifying common wiring issues through connection inspection, maintenance, and upgrades returns appliances to smooth function while avoiding hazardous defects.

CHAPTER 9

SAFETY PROTOCOLS AND PROCEDURES

Principles of Electrical Safety

Safe interaction with electrical systems relies on understanding key risks and implementing protective practices.
Insulation integrity minimizes shock risks. Damaged wire insulation exposed energized conductors. Frays, nicks, and cracks enable current to leak outside intended circuits, potentially through people. Maintaining insulation keeps the current contained.
Grounding creates a controlled path for stray currents to safely dissipate. Effective equipment grounding prevents voltage buildup on surfaces and cases. Proper bonding equalizes voltages and enables overcurrent devices to operate.
Circuit protection limits excessive currents. Fuses and breakers interrupt the flow from overloads or faults before wires overheat and ignite materials. Proper sizing ensures protection and coordination.
Physical barriers prevent contact with live parts. Enclosures and covers keep hands and objects away from exposed terminals and connections. Restricted access helps avoid accidental contact.
De-energizing equipment removes voltage during service. Lockout/tagout procedures isolate circuits and verify safe states. Testing before touching validates the absence of voltage.
Water safety employs GFCIs and separation from electrical systems in wet areas. Combining electricity and water can enable current to flow through people. Isolating them is imperative.
Adherence to fundamental safety guidelines reduces the likelihood of electrical dangers and diminishes the chances of experiencing shock, arc flash, electrocution, and fires. Developing safe work practices protects both personnel and equipment.

Implementing Effective Safety Protocols

Delivering consistently safe outcomes requires more than just knowledge of hazards and procedures. Organizations must work to instill safety as a cultural value and provide proper tools and training.
Leadership Commitment - Executives must fully embrace safety as a core priority, not just a compliance obligation. Resources for training, equipment, and accountability demonstrate the organization's values.
Hazard Assessments – Thorough initial and ongoing analysis informs protective measures. Establishing a base involves pinpointing potential hazards, determining their underlying causes, and devising strategies for their management..
Defined Policies and Procedures – Well-documented standards aligned with regulations and best practices remove ambiguity. Detailed protocols establish minimum expectations.
Training and Awareness – Broad delivery of engaging, practical training ensures comprehension of hazards and protocols. Tailored programs target roles and risks. Refreshers maintain readiness.
Protective Resources – Provide proper tools, systems, and equipment to enable compliance with procedures. Resources like PPE remove barriers to working safely.
Accountability – Observation, coaching, and discipline reinforce the sustainment of safe behaviors. Reward positive examples. Learn from incidents.
By instilling safety importance, equipping people properly, and setting clear expectations, organizations move beyond standard compliance to achieve a culture of openness, preparedness, and continuous learning in pursuit of injury-free workplaces.

Emergency Response Techniques

Being prepared to respond quickly and effectively in an emergency requires rigorous training, well-defined procedures, and the right mindset.
Responders should have extensive hands-on practice with likely emergency scenarios to build muscle memory and confidence under stress. Realistic, immersive simulations ingrain critical safety protocols.
Emergency action plans provide step-by-step guidance with checklists and decision flowcharts. Ready access to concise plans helps guide response. Prominent signage reinforces pivotal actions.
Coordination with stakeholders like medical providers establishes response interfaces. Unified command integrates multi-agency efforts on-site.

Fast access to response equipment like extinguishers, first aid kits, and PPE enables immediate action. Staging resources in strategic areas avoids delay.

Designated roles and responsibilities prevent confusion. Each responder must know their specific duties in a crisis. Cross-training expands capacity.

Controlled urgency balances speed with care under pressure. Remain focused without panic.

Continuous reassessment of hazards and progress guides, next actions. Maintain situational awareness.

Effective radio communication employs clear, concise language. Minimize chatter and confirm messages.

After each incident, debrief objectives, outcomes, and improvements. Revise protocols to mitigate recurring issues.

Organizations equip for swift and efficient emergency action to protect lives and assets by ingraining response capabilities via intensive practical training, establishing strong protocols, preparing equipment, and improving team collaboration.

The Role of Personal Protective Equipment

Electricians face an array of hazards from live circuits, falls, arc flash, and more. Proper selection and use of personal protective equipment is imperative to safety on the job.

Head Protection

- Hard hats protect against falling or fixed objects
- High voltage offers dielectric protection
- Replace after impact or degradation
- Chin straps for high winds
- Class E hard hats meet electrical testing

Eye Protection

- Safety glasses defend against debris
- Goggles seal against chemicals
- Face shields over glasses for arc flash
- Select appropriate shading when welding
- Replace scratched or damaged lenses

Hearing Protection

- Earplugs or earmuffs in high-noise areas
- Communication headsets in loud environments
- Routine noise monitoring

Hand Protection

- Leather, rubber, nitrile gloves for cuts, vibration
- Rubber insulating gloves rated for voltage
- Careful inspection for defects
- Proper leather protectors over insulating gloves

Foot Protection

- Steel or composite toe boots guard against falling objects
- Electrical hazard rated with non-conductive soles
- Slip-resistant soles
- Rubber boots for wet locations

Flame Resistant Clothing

- FR shirts, pants, and jackets for arc flash protection
- Appropriate HRC level for the hazard
- Long sleeves, shirts tucked in

Proper inspection, cleaning, maintenance, and training in PPE usage minimize injuries on the job. Employers must provide compliant equipment for electrical work hazards.

Creating a Culture of Safety on the Job

Fostering an organization-wide culture that consistently prioritizes safety requires more than policies and procedures. Leaders must demonstrate commitment through accountability, communication, training, and employee engagement. This establishes safety as a core value to guide decisions and behaviors.

Visible leadership commitment sets the tone from the top down. When executives engage in safety activities, inspect work areas, and allocate resources to protective measures, it underscores the importance of these initiatives. Leaders must be accountable for safety metrics and invest in training.

Open communication channels enable the reporting of hazards without reprisal. Discuss incidents thoroughly to extract lessons learned. Share safety alerts and measures widely. Welcomeemployee input and ideas for improving safety.

Comprehensive training ensures everyone understands risks, protocols, equipment, and emergency procedures. Hands-on practice builds skills. Tailor programs to roles and worksites. Consider language needs. Continue refreshers and new hire orientation.

Promote employee participation through safety committees that help develop best practices, conduct inspections, and share training. Recognize safe behaviors. Involve workers in incident reviews.

By demonstrating leadership commitment, fostering transparent communication, providing immersive training, and promoting employee involvement in safety programs, organizations can foster shared vigilance and responsibility for protecting people.

CHAPTER 10

NEC COMPLIANCE AND INSTALLATION REQUIREMENTS

Navigating the NEC: Key Articles and Sections

As the National Electrical Code has over 1000 pages and evolves through continuous revisions, a working knowledge of key articles and sections enhances interpretation and application. The following summary highlights important areas while not replacing the need for a deep study of code requirements.

Article 90 - Covers the scope, purpose, enforcement, and compliance requirements of the overall electrical code. Establishes a framework for proper interpretation and application of technical standards for safety.

Article 100 - Provides definitions for important terms and jargon used throughout the code requirements. Understanding precise NEC meanings of terminology is crucial for meeting code in specific contexts.

Article 110 - Identifies general requirements for electrical installations including circuit designations, conductor sizing, overcurrent protection coordination, equipment markings, listings, and minimum working clearances. Safety fundamentals.

Article 200 - Presents regulations for the branch circuit and feeder conductor sizing, overcurrent protection, identification, insulation, terminations, connections, and wiring methods. Ensures safe delivery of power.

Article 210 - Outlines branch circuit design standards including receptacle and lighting fixture distribution, GFCI and AFCI protection, voltage drop, conductor ampacity and derating, load calculations, and more. Enables compliant circuit layout.

Article 240 - Covers the proper selection and installation of overcurrent protective devices like fuses and circuit breakers. Necessary to prevent damage from excessive currents according to equipment ratings.

Article 250 - Specifies grounding and bonding requirements including system grounding, equipment grounding conductors, grounding electrode systems, and grounding arrangements. Enables effective fault current pathways.

Article 300 - Lists permitted wiring method types for power, lighting, and control circuits including conduits, tubing, cables, boxes, and insulation requirements for wet, damp, and dry locations. Matches wiring methods to environments for safety.

Article 310 - Provides conductor properties, current carrying capacity, temperature ratings, and insulation requirements. Ensures conductors are sized appropriately to prevent failure from overheating under expected loads.

Article 320 - Contains specific regulations about the permitted uses, conductor types, markings, splices, and installations of armored cable (type AC). Ensuring proper applications protects conductors.

Article 330 - Covers metallic conduits including rigid, intermediate, PVC-coated, and associated fittings. Lists permitted conductor fills, installation methods, and maintenance for robust conduit systems.

Article 338 - Addresses code requirements for service entrance cable or SE cable for above-ground outdoor feeders and branch circuits. Rules for support, protection, terminations, and splice limitations enable safe installations.

Article 340 - Specifies construction, permitted uses, ampacity, conductor counts, markings, and installation methods for underground feeder and branch circuit cables, both directly buried and in raceways. Protects underground cable circuits.

Article 342 - Provides standards for intermediate metal conduit (IMC). Includes trade sizes, northern threads, permitted fill tables, corrosion protection, crushed points, and installation methods for strength.

Article 344 - Contains code for rigid aluminum and aluminum alloy conduit. Lists rules for trade sizes, markings, fill capacity, join, fastener spacing, expansion fittings, and permitted uses for durability.

Article 348 - Addresses flexible metal conduit trade sizes, fittings, connectors, maximum length, grounding, and permit conditions and applications. Ensures flexibility without compromising safety and function.

Article 350 - Identifies code standards for liquid-tight flexible metal conduit construction, markings, fittings, connectors, installation, and permitted uses. Enables flexibility with liquid ingress protection.

Article 352 - Outlines specifications for rigid polyvinyl chloride (PVC) conduit products including schedule 80 PVC. Covers markings, joints, solvent cements, expansion fittings, bends, and application restrictions for use as a non-conductive conduit option.

Article 354 - Provides NEC standards for HDPE plastic ducts and conduits including material, pressure ratings, blockings, color coding, conductor protection, wall thickness, and applications both underground and above ground.

Article 358 - Contains code requirements for electrical metallic tubing (EMT) regarding trade sizes, corrosion protection, fill capacity, bends, mechanical strength, couplings, straps, markings, and uses permitted. Enables lightweight conduit option.

Article 366 - Specifies allowed cable tray systems and supports including welded wire mesh trays, solid bottom trays, ventilated troughs, ladder types, fill capacity, covers, and associated fittings. Facilitates organized cable runs.

Article 370 - Outlines standards for outlet, device, pull, and junction boxes including floor boxes. Covers volume calculations, required openings, conduit connections, supports, covers, grounding, nesting rules, and relationships with surrounding structural elements.

Article 374 - Provides NEC regulations for cell, communications, network-powered broadband, and data processing equipment enclosures including indoor and outdoor types. Ensures safe low-voltage distribution systems.

Article 382 - Defines nonmetallic extensions consisting of cores, tubes, and sheaths for securing and protecting wiring assemblies. Includes material and marking requirements for surface or recessed applications as a flexible raceway.

Article 384 - Addresses strut-type channel raceway materials, functional characteristics, permitted uses, fittings, covers, installation, grounding, conductor fill capacity, and sizing requirements. Accommodates wiring layout changes efficiently.

Article 388 - Covers surface metal raceways with a hollow elongated cross-section. Addresses multi-outlet assemblies, enclosing types, associated covers and fittings, installation, grounding, and fill capacity for wiring flexibility in commercial spaces.

Article 392 - Contains specifications for cable trays including materials, load support, securing cables, covers, fill calculations, ampacity adjustment factors, markings, grounding, and bonding. Safely manages large cable bundles.

Article 394 - Lists standards for conduit, tubing, and cable assemblies embedded in concrete like slab-on-grade foundations. Covers dimensions, conductor types, ampacity derating, rebar spacing, cover requirements, and permitted uses. Facilitates infrastructural power.

Article 396 - Addresses permitted applications of messengers that support open-run cables and wires. Includes strength, ampacity, clearances, guying, safety ties, and associated hardware specifications. Allows aerial cable runs.

Article 398 - Contains regulations for open wiring on insulators that route conductors along the surface of structures. Covers conductor types and ampacity, securing methods, locations, insulation, supports, and associated hardware. Alternative to conduit.

Article 400 - Provides general standards for flexible cords and cables regarding materials, current carrying capacity, markings, outer jacket properties, and permitted uses. Ensures safe implementation of flexible supply connections.

Article 404 - Specifies requirements for installed switches controlling lights and power circuits, including ratings, markings, access, grouping, enclosures, disconnects, and control devices such as dimmers. Essential for a proper switching setup..

Article 406 - Contains specifications for installed receptacles including ratings, configurations, listings, temperature ratings, grounding, GFCI protection, tamper-resistant types, and installation methods. Mandates safe outlet configurations.

Article 408 - Covers panelboard requirements including ratings, cabinets, overcurrent protection, markings, minimum wire bending space, clearance requirements, and branch circuit loads. Ensures properly configured and located electrical panels.

Article 409 - Specifies NEC standards for installed industrial control panels including markings, enclosures, terminals, conductor ampacity, disconnects, control, hazards like arc flash and short circuit, and associated testing. Necessary for safe performance.

Article 410 - Sets lighting fixture standards including temperature ratings, wiring connections, insulation, reflectors, recessed types, switches, and lamp holders. Ensures proper fixture selection and installation.

Article 411 - Addresses low voltage lighting standards [<1000V] like inflexible strip lighting under cabinets. Includes transformer types, locations, outputs, voltages, listings, and terminal temperatures. Regulates safe low-voltage lighting.

Article 422 - Details specifications for appliances including markings, disconnects, conductors, overcurrent protection, outlets, and equipment grounding. Ensures safe integration of appliances.

Article 424 - Provides standards for fixed electric space-heating equipment like baseboard heaters, duct heaters, and radiant heating sets. Covers installation, connections, controls, sizing, markings, and equipment listings. Enables code-compliant heating systems.

Article 430 - Specifies regulations for motors, motor branch circuits, motor overload protection, motor control circuits, motor controllers, and motor disconnects. Prevents damage and enables control over motors.

Article 440 - Addresses installation of air conditioning and refrigerating equipment. Includes conductors, overcurrent protection, equipment markings, disconnects, cord-and-plug items, and field-installed controllers. Ensures HVAC safety.

Article 450 - Covers transformers and transformer vaults including installation, ventilation, signs, conductor sizing, overload protection, grounding, markings, and locations. Protects this critical equipment.

Article 500-510 - Contains detailed classifications and electrical requirements for hazardous locations by zone, class, and group. Ensures ignition risks are minimized by division of areas, appropriate wiring methods, and equipment usage.

Article 520 - Provides standards for theaters, audience areas, and performance venues including stage switchboards, fixed stage equipment, portable stage equipment, grounding, and isolated grounding requirements. Reduces hazards involved in production.

Article 525 - Addresses electrical requirements for temporary installations like carnivals, circuses, fairs, and amusement parks. Includes generators, transformers, grounding, GFCI protection, wiring methods, and luminaires. Ensures safe short-term setups.

Article 530 - Specifies NEC rules for motion picture and television production studio sets, filming locations, equipment, wiring, overcurrent protection, GFCI, equipment grounding, and isolated grounding. Reduces risks associated with studio power distribution.

Article 540 - Contains regulations for motion picture projection rooms, film storage facilities, and multiple-theater complexes. Includes proper wiring, grounding, equipment, projection rooms, film storage, audience areas, wiring devices, listed equipment, HVAC, and exhaust systems to address safety for highly flammable films and chemicals.

Article 545 - Outlines electrical requirements for manufacturing facilities, covering motor and equipment controls, disconnects, conductor ampacity, adjustable speed drive systems, robotics equipment, and branch circuits for portable tools such as welders. Ensures safe industrial processes.

Article 547 - Covers agricultural electrical facility rules for areas like barns, grain processing equipment, crop storage, agricultural buildings, livestock handling equipment, fountains, and PTO-driven implements. Addresses wet, dust, and chemical corrosive locations common in agriculture as well as grounding requirements. Allows safe farming operations.

Article 550 - Contains standards for mobile homes and mobile home parks including service equipment, disconnecting means, grounding, wiring methods, GFCI protection, and receptacle requirements both in interior living spaces and for accessory buildings/structures. Establishes minimum safety requirements for residential mobility.

Article 551 - Identifies requirements for recreational vehicles and recreational vehicle parks such as wiring methods, receptacle ratings, disconnecting means, GFCI protection, equipment grounding, and bonding considerations. Reduces hazards associated with compact, mobile facility designs.

Article 552 - Provides regulations for electrical systems in park trailers which are larger recreational vehicles intended for extended occupancy. Includes power supply, grounding, wiring methods, GFCI protection, and equipment permitted for residential use in limited spaces.

Article 553 - Specifies standards for floating buildings on water like floating homes. Addresses shore power connections, wiring methods, grounding systems, and special equipment enclosures suitable for marine environments. Allows residences on the water while managing increased electrical hazards.

Article 555 - Covers regulations for marinas, boatyards, marine service facilities, wet slips, and yacht clubs. Includes shore power receptacles, wiring methods, ground fault protection, corrosion resistance, wiring devices, and hazardous locations considerations specific to wet, corrosive marine atmospheres.

Article 600 - Provides general requirements for electrical installations operating over 600 volts nominal. Covers conductor insulation, clearances, guarding, markings, cable trays, enclosures, and control equipment for high voltage. Necessary for safety.

Article 620 - Contains standards for elevators, escalators, lifts, and similar conveyances. Includes wiring, control systems, operating characteristics, grounding, mechanical devices, emergency signals, and remote maintenance diagnostics. Enables integration of extensive electrical transport systems.

Article 630 - Addresses requirements for electrical systems associated with anesthetizing locations in healthcare facilities. Includes isolated power, equipotential grounding, fixed and portable equipment, room ventilation, and gas delivery system connections. Mandates critical safety for anesthesiology.

Article 640 - Covers audio signal processing, amplification systems, and associated equipment for private and public venues. Includes cabling, grounding, bonding, power sources, raceways, and audio-electrical life safety/emergency systems. Ensures proper performance, and safety of specialized audio systems.

Article 645 - Provides standards for information technology systems including server rooms, telecommunications rooms, data centers, copper and fiber optic network cabling, grounding, bonding, fire protection, and HVAC requirements tailored to critical systems operations.

Article 680 - Contains regulations for swimming pools, fountains, spas, hot tubs, and similar installations. Includes transformers, lighting fixtures, receptacles, feeders, GFCI protection, underwater lighting, pool covers, bonding, and equipment locations for wet, high-exposure electrical equipment. Prevents shock/electrocution risks.

Article 690 - Specifies code requirements for solar photovoltaic power systems. Includes conductor sizing, overcurrent protection, disconnects, wiring methods, grounding, and connections to other sources. Ensures safe integration of renewable energy.

Article 695 - Addresses electrical standards for fire pump installations, wiring, protection against physical damage, feeders, listed equipment, power sources, and redundant pump controllers. Provides reliable power to mission-critical fire suppression systems.

Article 700 - Establishes NEC standards for legally required standby power systems in places of assembly or critical operations facilities. Includes universal systems, redundant supply options, transfer equipment, signals, listed equipment, maintenance, and capacity for backup power.
Article 701 - Provides regulations for optional standby backup power systems. Includes generators, transfer switches, transformers, uninterruptible power supplies (UPS), connections, protection, and load calculations for non-mandatory backup systems.
Article 702 - Contains design and installation requirements for optional energy storage systems like batteries, flywheels, and capacitors providing supplementary or backup power. Includes calculations, venting, spill control, disconnecting means, and safety.
Article 705 - Covers interconnected electric power production sources, utility-interactive inverters, transfer switches, interrupting devices, interconnection parameters, contracts, and maintenance considerations for distributed generation systems. Permits integration of local power generation.
Article 706 - Addresses critical operations power system (COPS) components that ensure continuous power to facilities like data centers that require uninterrupted service. Installation requirements facilitate redundancy and fail-safe capabilities.
Article 708 - Provides NEC standards for critical operations power system wiring and controls including transfer switches, life safety branch overcurrent protection, signals, listed equipment, maintenance, and capacity for facilities that mandate always-on power.
Article 725 - Contains regulations for remote control, signaling, and power-limited circuits that operate below 600V. Includes Class 1, 2, and 3 circuits for controlling equipment remotely, signaling systems, and communications critical to building operations like fire alarms and security systems.
Article 727 - Establishes requirements for instrumentation tray cable which transmits low-power signals in hazardous industrial locations. Includes construction, markings, ampacity, termination fittings, and installation methods for specialized cabling.
Article 760 - Specifies code for fire alarm systems, wiring methods, survivability, circuits, protection, and equipment. Includes power sources, control equipment, circuits, fire detectors, and emergency communication systems necessary for life safety and firefighting operations.
Article 770 - Provides standards for optical fiber cables and raceways. Contains specifications for cabling, cable marking, terminating components, grounding, installation, and testing. Ensures high-speed communications systems function properly.
Article 800 - Covers communications circuits requirements including grounding, separation from other systems, cabling, communications raceways, and equipment locations. Enables telephone, network, cable, radio, internet, and wireless systems to operate effectively.
Article 810 - Defines code standards for radio distribution systems like broadcast antennae and community wireless networks. Includes wiring, grounding, lightning protection, list equipment, and installation methods specific to wireless radio frequency distribution infrastructure.
Article 820 - Establishes NEC safety guidelines for community cable television and radio distribution systems. Specifies wiring, cabling, grounding, locations, power supply, and signal leakage limits to protect network infrastructure and users.
Article 830 - Provides safety regulations for network-powered broadband communications wiring. Contains powering limitations, separation from other systems, grounding, bonding, ultraviolet resistance, and indoor vs outdoor cable differences specific to network-powered communications.
Article 840 - Outlines requirements for distributing communications conductors and equipment within buildings. Includes pathways, spaces, grounding, mechanical execution, fire-stopping, and separation from other systems for occupancy safety and protection of communications capability.

NEC Requirements for Residential Installations

The National Electrical Code (NEC) establishes critical safety standards for residential dwellings to reduce risks of electrical fires, shock, and electrocution hazards. While compliance is legally mandated, knowledge of key requirements also guides homeowners and electricians in best practices for reliable, efficient systems. We will explore key NEC rules that underpin safe residential wiring.

At the service entrance, the main overcurrent protective device must match the ampacity of the service conductors and ratings of equipment. Service disconnects must also be readily accessible and located outside or immediately inside the nearest point of entrance. The service grounding electrode system must be installed per NEC regulations with proper ground rod quantity and spacing, suitable clamps, and exothermic weld connections. An equipment grounding conductor should connect to the grounding electrode and be included in all feeders and branch circuits to bond system enclosures.

Within the dwelling, sufficient lighting outlets and receptacle outlets must be installed in each habitable room, hallway, stairway, attached garage, and outdoor entrances/exits. Outlet spacing and placement requirements facilitate convenient usage without overloading individual circuits. Arc-fault circuit interrupter (AFCI) protection safeguards bedroom circuits to prevent fire ignition from electrical faults. Ground-fault circuit interrupter (GFCI) protection is mandated for bathrooms, kitchens, garages, outdoors, and other areas with shock risks from water exposure or damp environments.

Required branch circuit capacities and load calculations should incorporate the expected utilization of lighting, appliances, and equipment. Multi-wire branch circuits must share a common neutral and implement simultaneous disconnecting means. Box, conduit, and cable fill limits ensure no overcrowding of wiring which could cause dangerous overheating. Minimum bending radius rules protect conductors from damage. Reduced burial depths are allowed for underground conductors in shallow trenches under driveways or patios on residential property.

Key components of home electrical systems like service panels, switches, receptacles, and overcurrent devices must meet listing requirements for designated application use. Sturdy, well-fitted enclosures protect wiring and equipment. Proper terminations provide secure conductor connections. Installation meets minimum requirements, but often further improvements in safety and performance are wise investments for homeowners.

Additional recommendations enhance reliability and meet growing electricity demands in modern homes. Dedicated circuits isolate appliances with high-power drawers like air conditioners, electric vehicle chargers, and workshop tools. Arc-fault and ground-fault protections could be extended beyond NEC minimum circuits for increased safety. Added tamper-resistant receptacles to protect children. Surge protective devices buffer voltage fluctuations to protect equipment. Higher capacity service panels allow upgrades and expansions.

NEC Requirements for Commercial Installations

Constructing safe, code-compliant electrical systems in commercial spaces requires extensive expertise. The scale, complexity, and usage patterns of commercial facilities mandate rigorous adherence to National Electrical Code standards. We will examine key NEC rules that enable properly engineered electrical infrastructure to support building functions.

Service equipment must be robust with sufficient capacity, overcurrent protection ratings, fault tolerance, and physical security. Distribution systems utilize feeders, panelboards, and branch circuits to route power. Increased capacity requirements necessitate larger conductors and higher interrupt ratings. Load calculations incorporate all lighting, equipment, appliances, and expanded capacity for future needs.

Wiring methods must match designated building areas and uses. A rigid metal conduit provides robust mechanical strength. Electrical metallic tubing offers lighter-weight installations. Flexibility simplifies complex routing through cable trays, while liquid-tight flexible conduit withstands exposure to water or oils. Fire-rated construction preserves circuit integrity.

Emergency systems provide egress lighting, exit signs, fire detection, and alarms to maintain life safety when normal power fails. Legally required standby systems support critical loads like communications, fire pumps, and elevators. Optional standby systems back up HVAC, security, and similar facilities operations. Fuel cells, microgrids, and generators enable continuous power.

Hazardous location classifications segregate areas with explosion risks from flammable gasses, vapors, dust, or fibers. Wiring methods, equipment enclosures, and electrical ratings must prevent ignition. Workshops, paint booths, laboratories, and storage rooms with volatile materials require specialized electrical precautions.

Healthcare facilities incorporate technical requirements for patient care vicinities, equipment grounding, isolated power, anesthesia, and medical gas connections tailored to safely operate sensitive diagnostic and treatment functions.

Commercial kitchens, laundries, and baths necessitate GFCI protection against electric shock. Dedicated appliance branch circuits avoid overloads. Leak detection, ventilation, and temperature limits safeguard equipment like refrigerators, ranges, and water heaters. Strict separation between electrical and plumbing systems is mandated.

In large venues, NEC provisions enable lighting, audiovisual, and stage production capabilities while safeguarding occupants. Assembly points, exhibition halls, and studios integrate substantial electrical infrastructure with appropriate fail-safes and redundancies.

The NEC aims to establish minimum standards, but commercial facilities often benefit from additional upgrades. Arc-fault and tamper-resistant technologies enhance safety. Emergency management systems centralized monitoring. Power conditioning protects sensitive electronics. Communications cabling supports networks and devices. Lighting controls save energy.

With attentive electrical design, complex commercial structures can fulfill their purposes safely. nuanced NEC knowledge paired with engineering expertise enables holistic code compliance tailored to specialized commercial needs and innovations. When electrical systems are thoughtfully matched to building functions, they provide the safe, reliable infrastructure necessary to support operations.

Updates and Changes in the Latest NEC Edition

The National Electrical Code (NEC) evolves with each new edition to address emerging technologies, revised product standards, and lessons learned from field experience. The latest NEC provides significant updates that impact residential, commercial, and industrial electrical installations. Understanding these code changes enables compliance with safety standards and best practices.

Receptacles are now required in dwelling unit garages for electric vehicle charging. This facilitates the safe installation of EVSE equipment and avoids power cord hazards from overhead receptacles. Related cable ampacity rules have been added to accommodate increased loads from EV chargers. Licensed electricians must demonstrate familiarity with new EV charging requirements at the service panel, charging equipment, and dedicated branch circuits.

Arc fault circuit interrupter (AFCI) protections are newly mandated for additional dwelling unit circuits including laundry areas, utility rooms, basements, and crawl spaces. Expanded AFCI coverage improves electrical fire prevention in locations with appliances or where wiring may be subject to damage. Combination-type AFCIs mitigate both parallel and series arcs for total protection.

GFCI requirements have been extended to kitchen dishwasher branch circuits. Leakage current flowing through a dishwasher chassis can now be interrupted to prevent harm. Receptacles for countertop surfaces adjoining sinks must also be GFCI protected against water exposure when located within 6 feet of the outside sink edge.

In commercial facilities, new rules require receptacles in escalator machine rooms to facilitate safe maintenance. Storage battery systems over 50 volts must have disconnecting means for firefighter access during an emergency. New marking standards also identify equipment supplied by standby generators for easier recognition and manual transfer if necessary.

For swimming pools, pump motors now require GFCI protection when single-phase units larger than 15 amps are used. This provides shock hazard prevention for pool operators. Emergency shutoff switches for spas and hot tubs were also relocated to more readily accessible locations for quick access during a potential entrapment danger.

In industrial settings, new motor control and variable speed drive requirements help avoid drive system damage and nuisance tripping. Disconnecting means additions provide the ability to disconnect industrial machinery at the point of operation for servicing and emergency access. Control panels may also utilize zone selective interlocking when necessary to refine short-circuit isolation.

Solar photovoltaic systems have updated array ground fault detection needs along with revised requirements for module connection to grounded equipment. Rapid shutdown labeling provides critical visibility for first responders accessing rooftop solar arrays. Energy storage systems over 50 volts are also addressed, including disconnects, wire bending space, and charge controller connections.

NEC changes range from minor updates to substantial new requirements. Ongoing code familiarization allows electrical professionals to apply best practices and meet emerging needs as advancements are made. While adapting to changes takes concerted effort, enhancements made with each new edition ultimately improve safety and reliability if key revisions are understood and implemented properly.

Staying current with the latest NEC ensures electrical work meets the most rigorous and up-to-date safety standards. Technical knowledge plus code expertise empowers electrical crews to serve customers with legally compliant, highly effective system installations. The code development process assimilates field experiences to continuously refine requirements. Diligent NEC review and training prepare electrical professionals to apply new code knowledge correctly as updated editions are released.

CHAPTER 11

RENEWABLE ENERGY SYSTEMS

Introduction to Solar Power Systems

For many, the allure of solar power lies in its promise of harnessing a limitless, clean energy source. The sun, a constant in our skies, offers a sustainable way to meet our energy needs, and understanding how to tap into this resource is vital for modern electricians. Solar power systems, at their core, convert sunlight into electricity, providing an alternative to traditional fossil fuels. The journey into solar power begins with recognizing the fundamental components and operation of these systems.

Central to any solar power system is the photovoltaic (PV) cell, the device responsible for converting sunlight directly into electricity. These cells, made primarily from silicon, employ the photovoltaic effect—a process where light photons knock electrons loose from atoms, generating a flow of electricity. PV cells are typically grouped together to form panels or modules, which can be combined to create solar arrays. The efficiency of these cells, though dependent on various factors such as materials and design, has seen significant improvements over the years, making solar power a more viable option for widespread use.

Positioning and location play crucial roles in the effectiveness of solar power systems. For optimal performance, solar panels should be oriented to capture the maximum amount of sunlight, which often means facing them southward in the northern hemisphere and northward in the southern hemisphere. The angle of installation is also essential, ideally matching the latitude of the location to maximize exposure throughout the year. Shading, too, must be minimized, as even a small shadow can significantly reduce a panel's output.

Solar power systems can be designed for various scales and applications. Residential systems provide electricity for homes, often connected to the grid, allowing homeowners to sell excess power back to utility companies. Commercial systems, larger in scale, serve businesses and industrial facilities, sometimes as part of microgrids that enhance energy independence and resilience. Off-grid systems, meanwhile, operate independently, ideal for remote areas where traditional power infrastructure is absent.

The conversion of sunlight into usable electricity involves several critical components beyond the PV cells. Inverters are essential for converting the direct current (DC) generated by the panels into alternating current (AC), which is the form of electricity used by most household appliances and the grid. Inverters also play a role in maximizing power output through a process known as maximum power point tracking (MPPT), which adapts to changes in sunlight intensity.

Energy storage solutions, while not always necessary, are becoming increasingly common in solar power systems. Batteries store excess electricity produced during sunny periods, providing power during cloudy days or at night. Battery technology, particularly lithium-ion, has advanced significantly, offering greater storage capacity and longer lifespans. For those seeking energy independence or reliability in the face of power outages, integrating storage into solar systems can be highly beneficial.

The integration of solar power into existing electrical systems requires careful planning and expertise. Electricians must ensure that all components are compatible and that the system adheres to local codes and regulations. Proper wiring and grounding are crucial to prevent electrical faults, and protective devices, such as circuit breakers and surge protectors, safeguard both the system and the building it serves.

The economic advantages of solar power are substantial. While the initial investment in solar panels and installation can be significant, the long-term savings on electricity bills often justify the expense. Many regions also offer incentives, such as tax credits and rebates, to encourage the adoption of renewable energy. These financial benefits, along with the decreasing cost of solar technology, continue to drive the growth of solar installations globally.

Environmental considerations further bolster the case for solar power. Unlike fossil fuels, solar energy produces no greenhouse gases during operation, contributing to reduced carbon emissions and a cleaner atmosphere. By transitioning to solar, individuals and businesses can play a part in combating climate change and promoting sustainable practices.

For electricians venturing into the realm of solar power, staying informed about the latest technologies and industry trends is essential. Innovations in PV materials, such as perovskites, promise even greater efficiencies and cost reductions. Additionally, the integration of smart technology, allowing for remote monitoring and control of solar systems, is becoming increasingly prevalent.

Training and certification in solar power installation and maintenance are valuable assets for electricians seeking to enhance their skill set. Courses and programs designed to impart knowledge of solar technologies, system design, and safety

protocols can provide a competitive edge in the job market. As the demand for solar power continues to rise, qualified professionals will be in high demand to meet the needs of this growing industry.
While solar power systems offer numerous benefits, they are not without challenges. Variability in sunlight due to weather or geographic location can affect energy production, necessitating complementary energy sources or storage solutions to ensure a consistent power supply. Additionally, the disposal and recycling of solar panels, particularly as they reach the end of their lifespan, present environmental concerns that must be addressed.
Despite these challenges, the potential of solar power systems remains immense. As technology advances and economies of scale are realized, solar energy is poised to become an increasingly integral part of our energy landscape. For electricians, understanding the intricacies of solar power systems not only enhances their professional capabilities but also positions them at the forefront of a transformative energy revolution.

Wind Energy Basics and Applications

Harnessing the power of the wind has long been a pursuit of human ingenuity, from the ancient sails that propelled ships across the seas to the windmills that ground grain into flour. Today, wind energy stands as a cornerstone of the renewable energy revolution, offering a sustainable, low-emission alternative to fossil fuels. For electricians and energy professionals, understanding the basics and applications of wind energy is essential in the modern landscape of power generation.
Wind energy systems primarily revolve around the wind turbine, a device that converts the kinetic energy of moving air into mechanical power, which is then transformed into electricity. At the heart of a wind turbine is the rotor, featuring blades that capture wind energy. As the wind blows, these blades rotate, driving a shaft connected to a generator. This generator functions to convert mechanical energy into electrical energy, supplying power to homes, businesses, and grids.
The efficiency of a wind turbine is contingent on several factors, chief among them being the wind speed and the turbine's design. Wind speed is a critical determinant of energy output; turbines operate optimally at specific wind speeds, known as the rated wind speed. Below this, energy production diminishes, while above it, the turbine might shut down to prevent damage. The location of a turbine is therefore paramount, necessitating careful site selection to maximize exposure to consistent and strong winds.
Wind turbines come in various sizes and configurations, each suited to distinct applications. Large-scale turbines, often grouped into wind farms, are connected to national grids and contribute a significant portion of a region's electricity supply. These wind farms are typically situated in areas with favorable wind conditions, such as coastal regions or open plains. Offshore wind farms, positioned in bodies of water, capitalize on the stronger and more consistent winds found at sea, though they also present unique challenges in terms of installation and maintenance.
For smaller-scale applications, such as residential or community energy systems, smaller turbines are available. These turbines often complement other renewable energy sources, such as solar panels, to provide a balanced and reliable energy supply. In remote locations, where connecting to the grid is impractical or impossible, standalone wind energy systems offer an independent power solution, enhancing energy security and resilience.
The integration of wind energy into electrical systems requires a nuanced understanding of both its potential and limitations. Wind energy is inherently variable, with production fluctuating according to weather conditions and time of day. This variability necessitates the use of energy storage solutions or backup power systems, ensuring a steady supply of electricity even when the wind is not blowing. Moreover, grid operators must manage the intermittency of wind power, balancing supply and demand to maintain grid stability.
The environmental benefits of wind energy are substantial. Unlike fossil fuels, wind energy generates electricity without emitting greenhouse gases, contributing to reduced air pollution and a smaller carbon footprint. Additionally, wind turbines have a relatively low impact on land use, allowing for agriculture or other activities to continue around them. However, the installation and operation of turbines are not without ecological considerations. Potential impacts on wildlife, particularly birds and bats, necessitate careful planning and mitigation measures to minimize harm.
For electricians working with wind energy systems, the complexity of installation and maintenance cannot be understated. Constructing a wind turbine involves erecting towers, assembling rotors and nacelles, and connecting electrical components. Ensuring that all parts are securely fastened and properly aligned is critical to the turbine's performance and longevity. Regular maintenance, including inspections, lubrication, and component replacement, is essential to prevent mechanical failures and ensure optimal operation.
Safety is paramount in wind energy projects. The height of wind turbines and the mechanical nature of their components pose risks to workers during installation and maintenance. Adherence to safety protocols, including the use of personal protective equipment and adherence to guidelines, is necessary to safeguard personnel. Additionally, understanding the electrical aspects of wind energy systems, from wiring to grounding, is vital to preventing electrical hazards.

The economic landscape of wind energy is characterized by both opportunities and challenges. Wind energy projects often involve significant upfront costs, including site assessment, equipment purchase, and installation. However, the operational costs of wind energy are relatively low, as the wind itself is free and inexhaustible. Over time, the initial investment is offset by savings on electricity bills and potential revenue from selling excess power back to the grid. Incentives and subsidies, offered by governments to promote renewable energy adoption, further enhance the financial viability of wind energy systems.

For electricians and energy professionals, proficiency in wind energy technology can open doors to new career opportunities. As the demand for renewable energy continues to grow, skilled technicians are needed to design, install, and maintain wind energy systems. Training and certification programs tailored to wind energy provide the knowledge and skills required to excel in this field, covering topics such as turbine mechanics, electrical systems, and safety protocols.

The future of wind energy is promising, driven by technological advancements and a global commitment to sustainable energy solutions. Innovations in turbine design, such as taller towers and longer blades, are increasing the efficiency and capacity of wind energy systems. Developments in energy storage and grid integration are addressing the challenges of intermittency, enhancing the reliability of wind power. As these advancements continue, wind energy is poised to play an increasingly significant role in the global energy landscape.

Understanding wind energy basics and applications equips electricians with the tools to contribute to this evolving industry. By mastering the intricacies of wind energy systems, professionals can support the transition to a cleaner, more sustainable energy future.

Integrating Renewable Energy into Electrical Systems

Transitioning to renewable energy sources is not just a trend but a necessary evolution in how we generate and consume power. Integrating renewable energy into existing electrical systems requires a nuanced understanding of both traditional power infrastructure and the innovative technologies driving the shift toward sustainability. As electricians, mastering this integration is key to supporting a future that prioritizes clean energy solutions.

The integration process begins with understanding the types of renewable energy sources available and their respective characteristics. Solar and wind energy, for instance, are the most common, each with unique properties and challenges. Solar energy is predictable during daylight hours but requires storage solutions for nighttime or cloudy days. Wind energy, while more variable, can complement solar by providing power during different times or weather conditions. Hydroelectric power, although site-dependent, offers a more consistent energy supply, leveraging the kinetic energy of flowing water.

A critical aspect of integrating renewable energy involves connecting these systems to the existing electrical grid. This process, known as grid-tie, allows renewable energy systems to supply electricity directly to the grid, providing power to homes and businesses and potentially earning income through net metering. Net metering enables system owners to receive credit for excess energy produced, offsetting electricity costs. Ensuring a seamless grid connection requires careful coordination with utility companies and adherence to grid compatibility standards.

For an electrical system to efficiently incorporate renewable energy, inverters play a pivotal role. Inverters convert the direct current (DC) generated by renewable sources, such as solar panels, into alternating current (AC), which is the standard form of electricity used in homes and on the grid. Advanced inverters, often equipped with smart technology, not only perform this conversion but also optimize energy output through real-time monitoring and adjustments. These smart inverters can communicate with the grid, providing valuable data and enhancing grid stability.

Energy storage solutions are another vital component of renewable energy integration. Batteries store excess energy generated during peak production times, making it available when renewable sources are not actively producing power. Lithium-ion batteries are commonly used due to their high efficiency and long lifespan. For larger installations, such as commercial or community energy systems, more extensive storage solutions, including pumped hydro or compressed air, might be employed. Properly sizing and configuring these storage systems ensures that energy is available on demand, enhancing reliability and reducing dependence on the grid.

Microgrids represent a localized energy solution that integrates renewable sources into a smaller, self-sufficient electrical network. These systems can operate independently or in conjunction with the main grid, providing energy security and resilience, particularly in remote or disaster-prone areas. Microgrids often incorporate multiple renewable sources, along with storage and backup generators, to ensure a continuous power supply. Designing a microgrid involves careful planning and consideration of energy needs, resource availability, and potential growth.

The transition to renewable energy also involves retrofitting existing electrical systems to accommodate new technologies. This process can include upgrading wiring, panels, and protective devices to handle increased loads or different power qualities. Ensuring that all components are compatible and meet safety standards is crucial to prevent faults and hazards.

Electricians must be adept at troubleshooting and resolving any issues that arise during the integration process, using diagnostic tools and techniques to identify and correct problems efficiently.
Regulatory compliance is another critical aspect of integrating renewable energy. Local and national regulations govern the installation and operation of renewable energy systems, dictating requirements such as system size, placement, and safety standards. Navigating these regulations requires a thorough understanding of the applicable codes and the ability to work closely with regulatory bodies to secure necessary permits and approvals. Staying informed about changes in regulations, such as updates to the National Electrical Code (NEC), is essential for ensuring ongoing compliance and avoiding potential legal issues.
The economic implications of integrating renewable energy are multifaceted. While the initial investment in renewable technologies can be considerable, the long-term savings on energy costs often offset these expenses. Additionally, incentives such as tax credits, rebates, and grants are available to encourage the adoption of renewable energy, further enhancing the financial feasibility of these projects. For businesses and homeowners, the potential to generate income through energy sales or savings provides a compelling argument for transitioning to renewable sources.
As electricians work to integrate renewable energy into electrical systems, continuous education and training are vital. The rapid pace of technological advancement in the renewable energy sector necessitates staying current with new developments and best practices. Training programs and certifications, focused on renewable energy technologies and integration techniques, equip electricians with the skills needed to excel in this evolving field. Networking with industry professionals and participating in renewable energy forums and conferences can also provide valuable insights and opportunities for collaboration.
The benefits of integrating renewable energy extend beyond individual projects, contributing to broader environmental and societal goals. By reducing reliance on fossil fuels, renewable energy systems help decrease greenhouse gas emissions and mitigate climate change. The shift to renewables also promotes energy independence and security, reducing vulnerability to fluctuations in fossil fuel markets and geopolitical tensions. As more communities and industries embrace renewable energy, the collective impact on the environment and economy will be profound.
For electricians, the challenge and opportunity lie in facilitating this transition, ensuring that renewable energy systems are seamlessly integrated into existing electrical infrastructure. Mastery of renewable energy integration not only broadens professional expertise but also positions electricians as key players in the sustainable energy movement. By embracing the complexities and possibilities of renewable energy, electricians can drive innovation and support a future that prioritizes clean, reliable, and sustainable power for all.

Battery Storage Solutions

The evolution of renewable energy systems has brought with it a need for effective energy storage solutions. As solar panels and wind turbines generate electricity, the ability to store that energy for use during periods of low production is crucial. This is where battery storage solutions come into play, offering a way to stabilize energy supply, enhance grid reliability, and empower users with greater energy independence.
Batteries store energy in chemical form and convert it back to electrical energy when needed. The technology behind batteries has advanced significantly, with lithium-ion batteries leading the charge due to their high energy density, efficiency, and long cycle life. Unlike traditional lead-acid batteries, lithium-ion variants are lighter, require less maintenance, and can be discharged to a greater extent without affecting performance.
When considering battery storage solutions, one must understand the components and functionality of a battery storage system. At its core, a battery storage system consists of the batteries themselves, an inverter, and a battery management system (BMS). The BMS is crucial as it monitors and manages the charging and discharging processes, ensuring optimal performance and safety by preventing overcharging, deep discharging, and thermal runaway.
Sizing a battery storage system is an essential step in ensuring that energy needs are met without unnecessary expense. The size of the system depends on several factors, including the total energy consumption, the desired level of energy independence, and the availability of renewable energy sources. It's important to calculate the daily energy consumption in kilowatt-hours (kWh) and factor in the number of days of autonomy required, which refers to the number of days the system should operate without additional charging.
Battery storage systems can be configured in various ways to suit different applications. Grid-tied systems, for example, are connected to the main electricity grid and can store excess energy generated by renewable sources, which can then be used during peak demand times or during power outages. This configuration allows users to take advantage of net metering, where they can sell excess energy back to the grid, potentially reducing electricity bills.

Off-grid systems, on the other hand, operate independently of the main grid, making them ideal for remote locations where grid connection is not feasible. These systems rely heavily on battery storage to ensure a continuous power supply, often incorporating multiple renewable energy sources such as solar, wind, or hydro to maintain energy balance.

Hybrid systems combine both grid-tied and off-grid functionalities, offering the flexibility to store energy for personal use while still being able to feed excess power back into the grid. This configuration is particularly beneficial for those seeking energy independence without completely severing ties with the grid.

Maintenance and safety are paramount when dealing with battery storage solutions. Regular maintenance checks include inspecting battery terminals for corrosion, ensuring proper ventilation to prevent overheating, and monitoring the state of charge (SOC) to avoid deep discharges. The BMS plays a crucial role in maintaining battery health by providing real-time data on battery status and alerting users to potential issues.

Safety measures must also account for the possibility of thermal runaway, a condition where a battery cell's temperature rapidly increases, potentially leading to a fire. Proper installation, including adequate spacing, ventilation, and the use of fire-retardant materials, can mitigate these risks. Additionally, ensuring that all components are correctly rated for the system's voltage and capacity will prevent electrical faults and enhance overall safety.

The integration of battery storage solutions offers significant economic and environmental benefits. By storing excess renewable energy, users can reduce their reliance on fossil fuels, thereby decreasing carbon emissions and contributing to a cleaner environment. Financial incentives, such as tax credits and rebates, further enhance the cost-effectiveness of battery storage systems, making them an attractive investment for homeowners and businesses alike.

Emerging technologies continue to push the boundaries of battery storage capabilities. Solid-state batteries, for instance, promise even greater energy densities and safety improvements by replacing liquid electrolytes with solid materials. Flow batteries, which store energy in liquid electrolytes contained in external tanks, offer scalability and long cycle life, making them suitable for large-scale applications.

For electricians, understanding the intricacies of battery storage solutions is essential to effectively design, install, and maintain these systems. Training and certification programs focused on energy storage technologies provide the necessary knowledge and skills to excel in this growing field. By mastering battery storage solutions, electricians can offer clients cutting-edge energy solutions that align with the global shift toward sustainability.

As the demand for renewable energy grows, so too does the need for robust and reliable battery storage systems. By addressing the challenges of energy intermittency and enabling greater energy independence, battery storage solutions play a pivotal role in the transition to a more sustainable and resilient energy future. Electricians, as key enablers of this transition, have the opportunity to lead the charge in integrating innovative storage technologies that redefine how we harness and utilize renewable energy.

Regulatory Considerations for Renewable Installations

Navigating the intricate web of regulations is an essential aspect of installing renewable energy systems. As more individuals and businesses opt to harness the power of the sun, wind, and other renewable sources, understanding regulatory considerations becomes crucial. These regulations ensure safety, reliability, and efficiency, while also promoting the broader adoption of sustainable technologies.

At the core of regulatory considerations is compliance with local building codes and standards, which dictate the minimum requirements for the design, installation, and operation of renewable energy systems. These codes are established to ensure that installations are safe, functional, and in harmony with existing structures and utilities. Electricians must familiarize themselves with these codes, which can vary significantly from one jurisdiction to another. This involves understanding the specific requirements for system components, such as photovoltaic panels, wind turbines, inverters, and storage batteries.

Permitting is a critical step in the installation process, requiring electricians to secure the necessary approvals from local authorities before beginning work. Permits ensure that the proposed system meets all regulatory requirements and is compatible with the surrounding environment. The permitting process typically involves submitting detailed plans and specifications, which may require the expertise of licensed engineers or architects. Electricians must be prepared to address any questions or concerns that arise during this review process, demonstrating how the system complies with all applicable codes and standards.

Grid connection regulations are another key consideration, particularly for systems that feed electricity back into the public utility network. These regulations govern how renewable energy systems interact with the grid, ensuring safe and reliable operation. Electricians must understand the specific requirements for interconnection, which may include equipment standards, testing procedures, and contractual agreements with utility companies. Compliance with interconnection standards is essential to prevent issues such as power surges, voltage fluctuations, or grid instability.

Net metering policies are a vital aspect of regulatory considerations, providing a framework for how excess energy generated by renewable systems is credited or compensated. These policies allow system owners to receive financial credits for the surplus electricity they produce, which offsets their energy costs. The specifics of net metering, such as compensation rates and eligibility criteria, vary by jurisdiction, requiring electricians to be well-versed in the local regulations to advise clients accurately.

Safety standards are paramount in renewable energy installations, protecting both installers and end-users. The National Electrical Code (NEC) serves as a foundational document, providing guidelines for electrical safety in renewable energy systems. Electricians must adhere to NEC requirements, which cover aspects such as wiring methods, grounding, overcurrent protection, and equipment labeling. Compliance with these standards minimizes the risk of electrical hazards, ensuring the safe operation of the system throughout its lifespan.

Environmental regulations also play a significant role in renewable energy installations, particularly for large-scale projects. These regulations assess the potential impact of the installation on the surrounding environment, including considerations for land use, wildlife, and natural resources. Environmental assessments may be required to evaluate factors such as habitat disruption, noise pollution, and visual impact. Electricians working on such projects must collaborate with environmental experts to ensure that all regulatory requirements are met and that the project proceeds with minimal ecological disturbance.

Renewable energy incentives and subsidies are an important aspect of regulatory considerations, offering financial benefits to encourage the adoption of sustainable technologies. These incentives can take various forms, including tax credits, grants, rebates, and low-interest loans. Electricians should be knowledgeable about the available incentives in their area, helping clients understand and apply for these programs to reduce the overall cost of their renewable energy systems.

Training and certification requirements are another regulatory consideration, ensuring that electricians possess the necessary skills and knowledge to install renewable energy systems safely and effectively. Many jurisdictions require electricians to complete specialized training programs or obtain certifications specific to renewable technologies. These programs cover topics such as system design, installation techniques, safety protocols, and regulatory compliance. Electricians should seek out reputable training programs and stay current with continuing education opportunities to maintain their credentials and expand their expertise.

The dynamic nature of regulatory considerations means that electricians must stay informed about changes and updates to relevant codes, standards, and policies. This involves regularly reviewing industry publications, attending workshops and conferences, and participating in professional organizations. By staying abreast of regulatory developments, electricians can ensure that their installations remain compliant and that they continue to provide high-quality service to their clients.

The integration of renewable energy systems into the existing regulatory framework is a complex but essential task. By understanding and adhering to the myriad of regulations, electricians can facilitate the transition to sustainable energy sources while ensuring safety, reliability, and public trust. As the demand for renewable energy continues to grow, the role of electricians in navigating regulatory considerations becomes ever more critical, positioning them at the forefront of the sustainable energy movement.

CHAPTER 12

SMART HOME TECHNOLOGIES

Overview of Smart Home Devices

The evolution of technology has ushered in an era where homes are not just a place of shelter, but a hub of interconnected devices that enhance convenience, security, and energy efficiency. Smart home technologies have transformed the way we interact with our living spaces, offering a seamless integration of various systems that were once disparate. For electricians aiming to stay at the forefront of technological advancements, understanding smart home devices is imperative.

Smart home devices encompass a wide range of applications, each designed to improve daily living through automation and connectivity. At the heart of any smart home system is the network, often powered by Wi-Fi or other wireless protocols like Zigbee or Z-Wave. These protocols enable communication between devices, allowing them to function cohesively. For electricians, ensuring robust network connectivity is crucial for the effective operation of smart home technologies.

One of the most popular categories of smart home devices is smart lighting. These systems allow users to control lighting remotely, adjust brightness, and even change colors to suit different moods or occasions. Smart lighting can be scheduled to turn on or off at specific times, enhancing security by giving the appearance of occupancy. Electricians must be adept at installing compatible fixtures and ensuring that wiring and connectivity are optimized for these intelligent systems.

Smart thermostats are another cornerstone of home automation, offering precise control over heating and cooling systems. These devices learn user preferences and adjust settings automatically to improve comfort and reduce energy consumption. By integrating with other smart devices, such as sensors and weather stations, smart thermostats provide a comprehensive approach to climate control. Electricians play a key role in ensuring that these systems are properly installed and integrated with existing HVAC systems.

Security is a primary concern for many homeowners, and smart home devices offer advanced solutions in this area. Smart security systems include devices like cameras, doorbells, locks, and motion sensors, all of which can be monitored and controlled remotely. These systems provide real-time alerts and video feeds, allowing homeowners to keep an eye on their property from anywhere. Electricians must ensure that these devices are installed securely and that their connectivity is reliable to prevent unauthorized access.

The integration of voice assistants, such as Amazon Alexa, Google Assistant, and Apple Siri, has further enhanced the functionality of smart home devices. These assistants act as a centralized control point, enabling users to operate various devices using voice commands. Electricians must be familiar with the setup and configuration of these systems, ensuring that they are seamlessly integrated with the home's smart ecosystem.

Smart appliances bring another dimension to home automation, offering features that go beyond traditional functionality. From refrigerators that track inventory and suggest recipes to washing machines that optimize water usage, these appliances contribute to a more efficient and convenient household. Electricians need to understand the power requirements and connectivity options for these devices, ensuring they are installed correctly and safely.

Energy management is a significant benefit of smart home technologies, with devices designed to monitor and optimize energy consumption. Smart plugs and energy monitors provide insights into usage patterns, helping homeowners identify areas for improvement. By integrating with renewable energy sources, such as solar panels, smart home systems can further enhance energy efficiency. Electricians must be able to install these devices and provide guidance on how to maximize their benefits.

The concept of smart home technology extends beyond individual devices to encompass entire systems that operate harmoniously. Home automation platforms, such as SmartThings or HomeKit, serve as the backbone of a smart home, allowing for the integration and control of various devices from a single interface. Electricians must be skilled in setting up these platforms, ensuring compatibility and functionality across the entire system.

While the benefits of smart home technologies are numerous, they also present certain challenges. Network security is a paramount concern, as interconnected devices can be vulnerable to hacking if not properly secured. Electricians should educate homeowners on best practices for securing their networks, such as using strong passwords and enabling encryption.

Compatibility is another potential hurdle, as not all smart devices are designed to work together. Electricians must be knowledgeable about the various standards and protocols, ensuring that the devices they install can communicate effectively. This may involve recommending specific products or solutions that align with the homeowner's needs and existing systems.

The rapid pace of technological advancement means that electricians must commit to continuous learning to stay current with the latest smart home innovations. Training programs and certifications focused on home automation provide valuable

insights and skills, enabling electricians to offer cutting-edge solutions to their clients. Networking with industry professionals and participating in technology expos can also provide exposure to emerging trends and products.
Smart home technologies represent a significant shift in how we interact with our living environments, offering unparalleled convenience, security, and efficiency. For electricians, understanding and mastering these technologies is not just an opportunity but a necessity. By embracing the complexities and possibilities of smart home devices, electricians can enhance their service offerings and play a pivotal role in shaping the future of home automation.

Wiring and Networking for Smart Homes

Creating a smart home requires more than just purchasing the latest devices; it involves a comprehensive approach to wiring and networking to ensure seamless operation and connectivity. As an electrician, understanding the intricacies of wiring and networking is crucial to achieving a fully integrated smart home that meets clients' needs for convenience, security, and efficiency.
The foundation of any smart home is its electrical wiring. Traditional wiring systems need to be adapted to accommodate the demands of smart technologies, which often require additional circuits and loads. Planning for these demands begins with an assessment of the existing electrical infrastructure, identifying areas that require upgrades or alterations. This might involve installing dedicated circuits for high-demand devices, such as smart ovens or HVAC systems, to prevent circuit overloads and ensure reliable performance.
Structured wiring is a key element in smart home installations, providing a centralized location for all data and communication cables. This structured approach typically involves installing a panel or enclosure where all cables converge, making it easier to manage and troubleshoot the network. Electricians should plan the structured wiring layout meticulously, considering the placement of devices like routers, switches, and hubs, which will connect various smart devices throughout the home.
Cat6 or Cat6a Ethernet cables are often recommended for smart home networking, as they support high-speed data transmission and future-proof the installation against technological advancements. These cables provide a stable and reliable connection for devices that require consistent internet access, such as smart TVs, security cameras, and home automation hubs. Running these cables to key locations throughout the home ensures that all smart devices can connect to the network without relying solely on potentially unstable wireless connections.
While structured wiring is essential, wireless networking plays an equally significant role in modern smart homes. Wi-Fi networks provide the flexibility to connect mobile devices and smart gadgets without the need for extensive cabling. However, ensuring strong and consistent Wi-Fi coverage can be challenging, especially in larger homes or those with multiple levels. Strategic placement of wireless access points or mesh network systems can help eliminate dead zones and provide comprehensive coverage.
The integration of smart home devices often requires the use of different wireless protocols, such as Wi-Fi, Zigbee, or Z-Wave, each with its own advantages and considerations. Wi-Fi is ideal for high-bandwidth applications but can suffer from interference in crowded networks. Zigbee and Z-Wave, on the other hand, use mesh networking to allow devices to communicate with each other over longer distances, but may require additional hubs or bridges for compatibility with Wi-Fi. Electricians must understand these protocols to ensure that all devices can communicate effectively within the smart home ecosystem.
Another crucial aspect of smart home networking is ensuring network security. As smart homes become increasingly connected, they are also more vulnerable to cyber threats. Electricians should advise homeowners on best practices for securing their networks, such as changing default passwords, enabling encryption, and regularly updating firmware. Setting up a separate guest network for visitors can also help protect the primary network from unauthorized access.
Energy management is a significant benefit of smart home technologies, with devices and systems designed to monitor and optimize power consumption. Smart meters and energy management systems provide real-time data on electricity usage, allowing homeowners to make informed decisions about their energy consumption. Electricians should ensure that these systems are properly integrated with the home's electrical infrastructure and network, enabling seamless communication and data exchange.
The Internet of Things (IoT) is a driving force behind the proliferation of smart home technologies, connecting everyday objects to the internet for enhanced functionality and control. Electricians must be aware of how IoT devices interact with the home network, ensuring that they are configured correctly and do not cause interference or connectivity issues. This may involve setting up Quality of Service (QoS) settings on routers to prioritize traffic for critical devices.
The installation of smart home devices often requires collaboration with other professionals, such as IT specialists, to ensure that networking components are configured correctly. Electricians should be prepared to work closely with these

experts, sharing knowledge and insights to achieve the best possible outcome for the client. This collaborative approach can lead to more efficient installations and higher customer satisfaction.

Ongoing maintenance and support are vital for the long-term success of smart home installations. Electricians should offer clients guidance on maintaining their systems, such as regularly updating software and firmware, checking for network issues, and performing routine inspections of wiring and devices. Providing this support can help prevent problems before they occur and extend the lifespan of the smart home system.

As the demand for smart home technologies continues to grow, electricians have the opportunity to expand their skill sets and offer specialized services in this area. Training programs and certifications focused on smart home technologies provide valuable insights and skills, enabling electricians to design, install, and maintain these sophisticated systems. By staying informed about the latest advancements and best practices, electricians can position themselves as leaders in the smart home market.

Wiring and networking are the backbone of any smart home, enabling the seamless integration of devices and systems that enhance modern living. By understanding the complexities of these components, electricians can deliver installations that not only meet but exceed client expectations, paving the way for a future where smart homes are the norm rather than the exception.

Security Systems and Automation

The concept of home security has evolved dramatically with the advent of smart home technologies. No longer confined to simple locks and alarms, modern security systems are now sophisticated networks of interconnected devices, providing homeowners with unprecedented levels of control and peace of mind. For electricians, understanding the intricacies of these systems is essential in delivering installations that not only safeguard property but also enhance the overall living experience through automation.

At the forefront of smart security systems are smart cameras, which have become an integral part of home surveillance. These devices offer high-definition video capture, night vision, and motion detection, allowing homeowners to monitor their properties in real time from anywhere in the world. The installation of smart cameras requires meticulous attention to detail, ensuring that cameras are strategically placed to cover all vulnerable entry points without infringing on privacy. Electricians need to ensure robust network connections, as any lapse can compromise the efficacy of the surveillance system.

Smart doorbells have gained popularity for their ability to provide both security and convenience. Equipped with cameras and two-way audio, these devices allow residents to see and communicate with visitors at their door, even when they are not at home. Installation involves integrating the doorbell with the existing electrical system and network, which may require additional wiring or power sources, depending on the model. Electricians should advise homeowners on the best placement for these devices to ensure optimal field of view and audio clarity.

Security doesn't stop at cameras and doorbells; smart locks are another crucial component, offering keyless entry and remote access control. These locks can be programmed with unique entry codes for family members and guests, and can even be integrated with other smart home systems to trigger specific actions upon entry, such as adjusting lighting or thermostat settings. Electricians must be adept at installing these locks, ensuring that they are compatible with existing doors and that the electronic components are correctly wired and connected to the network.

Motion sensors and smart lighting play a vital role in deterring intruders and enhancing security through automation. Motion sensors can trigger alerts or activate smart lighting when movement is detected, providing both a deterrent and an immediate notification of potential intrusions. Integrating these systems requires electricians to strategically place sensors where they can detect unauthorized movement while minimizing false alarms from pets or other non-threatening sources.

Central to the effectiveness of smart security systems is the automation platform that ties all these devices together. Platforms such as SmartThings, HomeKit, or proprietary systems provided by security companies enable seamless integration and control of various devices. Electricians must be familiar with these platforms, ensuring that all components are configured to communicate effectively and that the system is user-friendly for the homeowner.

Automation is not limited to security alone; it extends to the overall management of the home environment. For instance, integrating security systems with smart lighting can create scenarios where lights turn on automatically when the home is entered, providing both convenience and energy efficiency. Similarly, connecting security systems to smart thermostats can adjust the home's climate control settings based on occupancy or time of day, enhancing comfort and reducing energy consumption.

Voice control, facilitated by virtual assistants like Amazon Alexa or Google Assistant, adds another layer of convenience to smart security systems. Homeowners can use voice commands to lock doors, view camera feeds, or arm the security system,

streamlining the interaction between user and technology. Electricians must ensure that these voice-controlled systems are properly configured and that the network connections are secure to prevent unauthorized access.
The integration of smart security systems requires a holistic approach, considering not only the installation of individual devices but also how they interact within the broader smart home ecosystem. Electricians should conduct thorough assessments of the home's layout and existing infrastructure to identify potential challenges and opportunities for integration. This may involve upgrading network capabilities or adding additional power sources to support new devices.
Security is paramount, and protecting the network that connects all smart home devices is a critical step in safeguarding the home. Electricians should educate homeowners on securing their networks, such as setting strong, unique passwords for all devices and enabling network encryption. Regular updates to device firmware are also crucial, as they often include security patches that protect against vulnerabilities.
Emerging technologies continue to shape the future of smart security systems. Advances in artificial intelligence and machine learning are leading to more sophisticated analytical capabilities, enabling systems to differentiate between genuine threats and non-threatening activities. Electricians should stay informed about these developments, as they may present new opportunities and considerations in future installations.
For homeowners, the benefits of smart security systems extend beyond protection to include increased property value and peace of mind. Electricians can play a pivotal role in guiding clients through the selection and installation process, ensuring that their security systems are tailored to their specific needs and preferences. By providing expert advice and support, electricians can foster trust and confidence, positioning themselves as indispensable partners in the journey toward a smarter, more secure home.
Understanding the intricacies of security systems and automation is essential for electricians who wish to excel in the smart home sector. By mastering the installation and integration of these technologies, electricians can deliver solutions that provide homeowners with unmatched control, convenience, and security, redefining what it means to feel safe and comfortable in one's own home.

Energy Management in Smart Homes

Harnessing the potential of smart home technologies to optimize energy consumption is a game-changer in today's world. As energy costs rise and environmental concerns become more pressing, homeowners are increasingly turning to smart solutions to manage their energy use more efficiently. For electricians, understanding how to implement and configure these energy management systems is crucial in providing clients with the tools they need to reduce their carbon footprint and save on utility bills.
At the heart of energy management in smart homes is the smart meter, a device that provides real-time data on energy consumption. Unlike traditional meters, smart meters offer detailed insights into how and when energy is used, enabling homeowners to make informed decisions about their consumption habits. Electricians must be adept at installing these devices, ensuring they are properly integrated with the home's electrical system and connected to the network for remote monitoring and control.
Smart thermostats are a pivotal component of energy-efficient homes, allowing precise control over heating and cooling systems. By learning user preferences and adjusting settings automatically, these devices optimize climate control for both comfort and efficiency. Electricians play a key role in installing and configuring smart thermostats, ensuring they are compatible with existing HVAC systems and connected to the home's network for seamless operation.
Lighting is another area where smart technologies can significantly impact energy consumption. Smart lighting systems allow users to control lights remotely, set schedules, and adjust brightness based on occupancy and natural light conditions. This flexibility reduces unnecessary energy use and can be particularly effective when integrated with motion sensors. Electricians need to ensure that smart lighting systems are correctly installed, with compatibility across different devices and platforms for cohesive operation.
Smart plugs and power strips offer a straightforward way to manage energy use by controlling the power supply to individual appliances and devices. These devices can be programmed to turn off during periods of inactivity or when energy prices are at their peak, contributing to overall energy savings. Electricians should ensure that these smart solutions are installed in locations where they can effectively manage the power flow to high-energy devices, such as entertainment systems and kitchen appliances.
Renewable energy sources, such as solar panels, are increasingly being integrated into smart home systems to further enhance energy efficiency. By generating their own power, homeowners can reduce reliance on grid electricity and lower their energy bills. Electricians must be knowledgeable about the installation and integration of renewable energy systems, ensuring they work harmoniously with smart home technologies to optimize energy use.

Energy management systems (EMS) provide a centralized platform for monitoring and controlling all aspects of a home's energy use. These systems offer a comprehensive view of consumption patterns, helping homeowners identify areas for improvement and track the effectiveness of their energy-saving measures. Electricians play a crucial role in setting up these systems, ensuring they are configured correctly and that all devices are connected for accurate data collection and analysis.

The integration of energy management with home automation platforms allows for sophisticated control over energy use. By connecting devices such as smart thermostats, lighting, and appliances to a central hub, homeowners can create automated routines that adjust settings based on time of day, occupancy, or even weather conditions. Electricians must ensure that these automation systems are set up to communicate effectively, providing a seamless user experience.

One of the significant benefits of smart energy management is demand response, a feature that adjusts energy consumption based on signals from the utility company. During peak demand periods, smart devices can reduce their energy use, helping to balance the load on the grid and potentially earning financial incentives for the homeowner. Electricians should be familiar with how demand response programs work and how to configure smart devices to participate in these programs.

Security and privacy are important considerations in energy management systems, as they involve the transmission of sensitive data about a homeowner's energy use. Electricians should educate clients on securing their networks and devices, ensuring that personal information is protected from unauthorized access. This may involve setting strong passwords, enabling encryption, and regularly updating firmware to protect against vulnerabilities.

Looking to the future, advances in technology continue to drive innovation in energy management. Machine learning algorithms are being developed to provide even greater insights into energy use, predicting consumption patterns and recommending personalized energy-saving strategies. Electricians should stay informed about these developments, as they offer new opportunities to enhance the efficiency and effectiveness of smart home systems.

For homeowners, the benefits of smart energy management extend beyond cost savings to include increased comfort, convenience, and environmental responsibility. By optimizing their energy use, they can contribute to a more sustainable future while enjoying the many advantages of a smart home. Electricians, as key enablers of this transformation, have the opportunity to lead the way in helping clients achieve their energy goals.

Understanding the intricacies of energy management in smart homes is essential for electricians who wish to offer cutting-edge solutions to their clients. By mastering the installation and configuration of these systems, electricians can deliver installations that provide homeowners with the tools they need to take control of their energy use, paving the way for a more sustainable and efficient living environment.

Troubleshooting Smart Home Technologies

The rapid advancement of smart home technologies has transformed modern living, offering unparalleled convenience and control. However, these sophisticated systems are not without their challenges. Troubleshooting is an essential skill for electricians, ensuring that installations run smoothly and that any issues are swiftly resolved. This chapter delves into the common problems encountered with smart home technologies and provides practical solutions to address them.

Connectivity issues are among the most frequent problems faced in smart home setups. Devices rely heavily on stable network connections to function correctly, and any disruption can lead to malfunction. The first step in resolving connectivity problems is to check the network infrastructure. Ensure that routers and access points are positioned optimally to provide comprehensive coverage throughout the home. Interference from other electronic devices or physical obstacles like walls and furniture can weaken Wi-Fi signals. Repositioning these devices or adding range extenders can help improve connectivity.

Another frequent issue is device compatibility. With various manufacturers and protocols, not all smart devices are designed to work seamlessly together. Electricians should verify that devices are compatible with the existing smart home ecosystem before installation. If compatibility issues arise, it may be necessary to update firmware or use a hub or bridge to facilitate communication between disparate devices. Keeping abreast of the latest updates and compatibility notes from manufacturers can prevent these issues from arising.

Power supply problems can also affect the performance of smart home devices. Devices may not function correctly if they do not receive sufficient power or if there are fluctuations in the power supply. Electricians should ensure that all devices are connected to suitable power sources and that wiring is up to code. In some cases, installing surge protectors or uninterrupted power supplies (UPS) can prevent voltage spikes or power outages from disrupting device operation.

Software glitches are not uncommon in smart home systems, often manifesting as devices freezing or failing to respond to commands. Rebooting the device or performing a factory reset can often resolve these issues. However, it's crucial to back up settings and configurations before performing a factory reset to avoid losing important data. Regularly updating device software and firmware can also help prevent glitches by ensuring that devices operate with the latest improvements and security patches.

User error is another common source of problems in smart home systems. Complex interfaces or misunderstanding of device capabilities can lead to perceived malfunctions. Electricians should take the time to educate homeowners on how to use their smart home technologies effectively. Providing clear instructions and demonstrations can empower users to troubleshoot minor issues independently. Additionally, creating a user-friendly guide or cheat sheet with common troubleshooting steps can be a valuable resource for homeowners.

Security concerns are an ever-present challenge in smart homes. If devices are not properly secured, they can be vulnerable to hacking or unauthorized access. Electricians should advise homeowners on best practices for securing their smart home systems. This includes setting strong passwords, enabling two-factor authentication, and regularly updating security settings. Ensuring that all devices are on a secure network and that guest networks are used for visitors can also enhance security.

Occasionally, hardware failures can occur, necessitating replacement or repair of the affected device. Electricians should have a reliable process for diagnosing hardware issues, which may involve checking for physical damage, testing connections, or swapping out components. In cases where a device needs to be replaced, it's important to verify that the new device is compatible with the existing system and that it is configured correctly.

Integration problems can arise when adding new devices to an existing smart home system. The seamless operation often requires careful configuration and synchronization of devices. Electricians should ensure that all new devices are added to the central automation platform and that any necessary permissions or settings are adjusted to allow proper integration. Testing the system thoroughly after adding new devices can help identify any integration issues early on.

Latency issues, where commands are delayed or devices respond slowly, can be particularly frustrating for homeowners. These issues are often related to network congestion or insufficient bandwidth. Electricians can address latency by optimizing the network, which may involve upgrading the router, increasing internet bandwidth, or implementing Quality of Service (QoS) settings to prioritize traffic for critical devices.

The importance of regular maintenance and monitoring cannot be overstated. Proactive maintenance can prevent many common issues from developing and can extend the lifespan of smart home devices. Electricians should advise homeowners on routine checks, such as verifying network connections, updating software, and inspecting physical installations for wear or damage.

Electricians should also be prepared to offer support and troubleshooting services as part of their business model. Providing ongoing support builds trust with clients and encourages long-term relationships. Developing a feedback loop with clients can also provide valuable insights into common issues and areas for improvement in smart home installations.

In the rapidly evolving world of smart home technologies, troubleshooting is a critical skill that ensures systems operate smoothly and meet homeowners' expectations. By understanding the common problems and implementing effective solutions, electricians can deliver reliable and high-performing smart home systems that enhance modern living. This commitment to excellence not only satisfies clients but also positions electricians as trusted experts in the field of smart home technology.

CHAPTER 13
ADVANCED TROUBLESHOOTING TECHNIQUES

Systematic Approach to Fault Finding

Navigating the world of smart home technologies demands not only technical proficiency but also a methodical approach to identifying and resolving faults. A systematic strategy for fault finding is essential for electricians striving to maintain the integrity and functionality of these intricate systems. This chapter outlines a comprehensive framework for diagnosing and addressing issues in smart home technologies, drawing on practical insights and a structured methodology.

Understanding the system as a whole is the first step in effective fault finding. Smart home technologies are complex networks of interconnected devices, each with its own function and dependencies. Electricians must familiarize themselves with the specific layout and configuration of the system they are working on. This involves reviewing installation documents, schematics, and any user manuals available to gain a clear picture of how the system is supposed to operate.

Once the system is understood, the next step is observation. Carefully observe the symptoms reported by the homeowner or noticed during routine checks. These symptoms are often the key to identifying the underlying issue. Whether it's a device failing to respond, intermittent connectivity problems, or unexpected behavior, every detail matters. Electricians should take note of when the issue occurs, which devices are affected, and any recent changes to the system that might have triggered the problem.

The process of elimination is a powerful tool in fault finding. Start by isolating each component to determine if it is the source of the problem. This can be achieved by systematically testing each device independently. If a device functions correctly in isolation but not when connected to the system, the issue may lie in the integration or network configuration. Conversely, if a device fails to operate even when isolated, it may indicate a problem with the device itself or its power supply.

Electrical faults are a common source of issues in smart home technologies. Checking the power supply is a fundamental step in the fault-finding process. Ensure that all devices are receiving the correct voltage and that there are no loose connections or damaged wires. Electricians should use multimeters to measure voltage levels and continuity, providing valuable data on the health of the electrical system. Pay particular attention to power-hungry devices that may be overloading circuits or causing voltage drops.

Network issues are another frequent culprit in smart home faults. Verify that all devices are connected to the network and that there are no IP conflicts or signal interference. Checking the router and any access points for proper configuration is crucial. Electricians should also ensure that firmware is up to date, as outdated software can lead to compatibility issues and security vulnerabilities. Testing network speed and reliability can help identify if bandwidth limitations are affecting device performance.

Software glitches and configuration errors can often be resolved through simple troubleshooting steps. Restarting devices or performing a factory reset can clear temporary software issues, although care should be taken to back up settings first. Electricians should ensure that all devices are running the latest firmware, as manufacturers frequently release updates to fix bugs and improve performance. Reconfiguring device settings or reinstalling apps and integrations may also resolve persistent software-related problems.

Communication is a critical component of effective fault finding. Electricians should maintain open lines of communication with homeowners, gathering as much information as possible about the issue and keeping them informed about the steps being taken to resolve it. This not only helps in diagnosing the problem but also builds trust and confidence with clients. Encouraging homeowners to report any unusual behavior promptly can prevent minor issues from escalating into major faults.

Documentation is an often-overlooked aspect of fault finding, yet it is invaluable for both immediate troubleshooting and future reference. Electricians should keep detailed records of the system configuration, any changes made, and the steps taken during the fault-finding process. This documentation can serve as a roadmap for resolving similar issues in the future and can be a valuable resource when training new team members or collaborating with other professionals.

When dealing with complex faults, collaboration with other experts may be necessary. Electricians should not hesitate to seek assistance from IT specialists, device manufacturers, or other professionals with specific expertise. This collaborative approach can provide new perspectives and solutions that might not have been considered otherwise, leading to more effective problem resolution.

Preventative measures are equally important in minimizing the occurrence of faults in smart home systems. Regular maintenance checks, including testing network connections, inspecting wiring, and updating software, can help identify

potential issues before they become significant problems. Electricians should also provide homeowners with guidance on how to maintain their systems, such as recommending regular software updates and advising on optimal device placement to avoid network interference.
As smart home technologies continue to evolve, so too do the challenges associated with fault finding. Electricians must stay informed about the latest developments in technology and best practices for troubleshooting. Continuous learning and professional development are essential for maintaining expertise and ensuring that electricians can provide the highest level of service to their clients.
A systematic approach to fault finding is essential for electricians working with smart home technologies. By understanding the system, observing symptoms, isolating components, and leveraging a range of diagnostic tools, electricians can efficiently identify and resolve issues. This methodical approach not only restores functionality but also enhances the overall reliability and performance of smart home systems, ensuring that homeowners can enjoy the full benefits of their technologically advanced environments.

Using Diagnostic Tools and Equipment

Effectively navigating the realm of smart home technologies requires not only a deep understanding of the systems but also proficiency in using diagnostic tools and equipment. For electricians, mastering these tools is crucial to diagnosing issues, ensuring optimal performance, and maintaining the integrity of complex installations. This chapter provides a detailed guide to the essential diagnostic tools and equipment that every electrician should be familiar with when working with smart home technologies.
The multimeter is an indispensable tool in any electrician's arsenal, offering the ability to measure voltage, current, and resistance with precision. When dealing with smart home devices, verifying the electrical parameters is often the first step in troubleshooting. A multimeter can help identify issues such as inadequate power supply or circuit continuity problems, providing quick and accurate readings that guide further investigation.
Circuit testers are another vital piece of equipment, used to determine whether electrical circuits are live and functioning correctly. These testers come in various forms, from non-contact voltage testers that provide a quick check of live wires to more advanced plug-in testers that can diagnose wiring faults in outlets. Electricians should utilize circuit testers to ensure that all circuits are correctly wired and that there are no hidden faults that could affect the performance of smart home devices.
Network analyzers have become increasingly important as smart home technologies rely heavily on stable and robust network connectivity. These tools help electricians assess the health of a network, identifying issues such as signal interference, bandwidth limitations, or IP conflicts. By providing detailed insights into network performance, network analyzers enable electricians to make informed adjustments to optimize connectivity for smart devices.
Thermal imaging cameras offer a unique perspective by detecting heat patterns and highlighting temperature variations in electrical systems. These cameras are particularly useful for identifying overheating components or loose connections that may not be visible to the naked eye. By pinpointing hot spots, electricians can address potential safety hazards before they lead to system failures or damage.
Cable testers are essential for verifying the integrity of data and power cables used in smart home installations. These devices test for continuity, shorts, and wiring faults, ensuring that all connections are secure and functioning as intended. Electricians should regularly use cable testers to confirm that both new installations and existing systems are free from cable-related issues that could disrupt performance.
Oscilloscopes, though more advanced, provide valuable insights into the electrical signals within a smart home system. By visualizing waveforms, oscilloscopes allow electricians to analyze the behavior of electrical signals and identify anomalies that may indicate underlying issues. While not always necessary for basic troubleshooting, oscilloscopes are invaluable for diagnosing complex or intermittent problems that require a deeper understanding of electrical behavior.
Signal generators are used to inject known signals into a system, allowing electricians to test the response of various components. These tools are particularly useful when diagnosing communication issues between smart devices, as they can help isolate the source of a problem by simulating different scenarios. Electricians can use signal generators to verify that devices are responding correctly to input signals and that communication pathways are functioning as expected.
The use of diagnostic software and apps has become increasingly prevalent with the rise of smart home technologies. Many manufacturers provide proprietary apps that offer diagnostic capabilities, allowing electricians to troubleshoot devices remotely or perform updates. These apps can provide real-time data on device performance, error logs, and configuration settings, offering a comprehensive overview of the system's health. Electricians should familiarize themselves with the available software tools for the devices they work with, leveraging these resources to streamline diagnostics and maintenance.

Documentation tools, such as digital cameras or mobile apps, are invaluable for recording and organizing information during diagnostics. By taking photos or notes, electricians can document the layout and configuration of a system, track changes made, and create a reference for future troubleshooting. This documentation not only aids in immediate diagnostics but also serves as a valuable resource for ongoing maintenance and client communication.

Safety equipment, while not diagnostic in the traditional sense, is essential when working with electrical systems. Electricians should always use insulated gloves, goggles, and other protective gear to ensure their safety while diagnosing faults. Additionally, lockout/tagout devices should be employed to prevent accidental energizing of circuits during diagnostics, safeguarding both the electrician and the system.

Calibration tools are necessary to ensure that diagnostic equipment remains accurate and reliable. Regular calibration checks are essential for maintaining the precision of tools such as multimeters, network analyzers, and oscilloscopes. Electricians should adhere to manufacturers' recommendations for calibration intervals and procedures, ensuring that their equipment remains in optimal condition for accurate diagnostics.

Training and continuous learning are crucial in keeping abreast of the latest diagnostic tools and techniques. As smart home technologies evolve, new diagnostic equipment and methodologies are developed to address emerging challenges. Electricians should engage in ongoing professional development, attending workshops, courses, or seminars to stay informed about the latest advancements in the field.

By mastering the use of diagnostic tools and equipment, electricians can efficiently troubleshoot and maintain smart home systems, ensuring that they operate at peak performance. This expertise not only enhances the reliability of installations but also positions electricians as trusted professionals capable of addressing the complex challenges of modern smart home technologies. Through a combination of technical skill, detailed diagnostics, and a commitment to excellence, electricians can deliver superior service and support to their clients, fostering confidence and satisfaction in their smart home solutions.

Common Issues in Complex Circuits

As the landscape of home technology evolves, the complexity of electrical circuits within smart homes continues to grow. Electricians must be adept at identifying and resolving the myriad issues that can arise in these sophisticated systems. Understanding the common problems associated with complex circuits is crucial for ensuring the seamless operation of smart home technologies and maintaining customer satisfaction.

One of the most prevalent issues in complex circuits is overloading. Smart homes are often equipped with a multitude of devices, each drawing power from the electrical system. When too many devices are connected to a single circuit, the demand for electricity can exceed the circuit's capacity, leading to tripped breakers or blown fuses. Electricians should assess the load requirements of each circuit, ensuring that it is adequately rated to handle the devices it supports. If overloading is a frequent issue, it may be necessary to redistribute devices across multiple circuits or upgrade the electrical panel to accommodate increased demand.

Loose connections are another common problem that can disrupt the functionality of complex circuits. Over time, connections can become loose due to thermal expansion and contraction, vibration, or improper installation. These loose connections can lead to intermittent power loss, arcing, or overheating, posing both operational and safety risks. Electricians should regularly inspect connections, ensuring that all terminals are secure and that wires are properly seated. Using a torque screwdriver to apply the correct amount of pressure can help maintain secure connections and prevent future issues.

Voltage drops can significantly impact the performance of smart home devices, especially those located far from the main power source. When the voltage supplied to a device is lower than its rated requirement, it may not function correctly or at all. Voltage drops are often caused by long cable runs, undersized conductors, or high-resistance connections. Electricians should calculate the voltage drop for each circuit, using appropriately sized conductors to minimize losses. If voltage drop is a persistent issue, installing step-up transformers or relocating devices closer to the power source may be necessary.

Harmonics are a less commonly understood issue in complex circuits but can have a profound effect on the performance and longevity of electrical systems. Harmonics are distorted electrical waveforms that result from the operation of non-linear loads, such as LEDs, computers, and other electronic devices commonly found in smart homes. These distorted waveforms can cause overheating, equipment malfunction, and increased energy consumption. Electricians should use harmonic analyzers to identify and measure harmonics in the system. Installing filters or specialized transformers can mitigate the impact of harmonics, ensuring the smooth operation of smart home technologies.

Grounding issues are critical to the safety and functionality of any electrical system, and smart homes are no exception. Improper grounding can lead to dangerous situations, such as electric shock or equipment damage. Ground loops, where multiple paths to ground exist, can cause interference with sensitive electronic devices. Electricians should verify that all circuits are properly grounded, using ground testers to ensure low-resistance connections. Correcting grounding issues often involves reconfiguring the grounding system or installing isolation transformers to eliminate ground loops.

Interference and noise are common issues in circuits that support smart home technologies. Devices such as wireless routers, Bluetooth speakers, and even fluorescent lighting can introduce electromagnetic interference (EMI) or radio frequency interference (RFI) into the system. This interference can disrupt the operation of sensitive devices, leading to erratic behavior or communication failures. Electricians should identify sources of interference and take steps to shield or filter affected circuits. Twisted pair cables, ferrite beads, and EMI filters are effective tools for minimizing interference and ensuring reliable device performance.

Short circuits and ground faults are potentially dangerous issues that can lead to equipment damage, fire, or personal injury. Short circuits occur when a low-resistance path is created, allowing excessive current to flow through the circuit. Ground faults are similar but involve current leaking to ground. Both issues can result from damaged insulation, faulty wiring, or equipment failure. Electricians should use insulation resistance testers and ground fault circuit interrupters (GFCIs) to detect and address these problems. Promptly repairing damaged wiring and replacing faulty equipment is essential to maintaining a safe and functional electrical system.

Thermal issues, including overheating, are a concern in complex circuits with high power demands. Excessive heat can degrade insulation, damage components, and increase the risk of fire. Electricians should monitor the temperature of critical components using thermal imaging cameras, identifying hot spots that may indicate underlying issues. Ensuring adequate ventilation, using heat sinks, or upgrading components to higher-rated versions can help mitigate thermal issues and extend the lifespan of the electrical system.

Compatibility issues can arise when integrating new devices into an existing smart home system. Mismatched voltages, incorrect wiring configurations, or incompatible communication protocols can prevent devices from operating correctly. Electricians should verify that new devices are compatible with the existing system, consulting manufacturer specifications and documentation. When compatibility issues arise, solutions may include reconfiguring the wiring, installing adapters, or updating software and firmware to support new protocols.

The rise of smart home technologies has introduced new challenges in managing and maintaining complex circuits. However, by understanding and addressing the common issues that arise, electricians can ensure that smart home systems operate safely and efficiently. Through careful planning, regular maintenance, and the application of best practices, electricians can deliver reliable and high-performing solutions that meet the demands of modern living. This expertise not only enhances the value of smart home installations but also fosters trust and satisfaction among homeowners, positioning electricians as indispensable partners in the journey toward smarter, more connected homes.

Case Studies in Troubleshooting

The journey to mastering advanced troubleshooting techniques involves not just theoretical knowledge, but also the practical application of skills in real-world scenarios. Through a series of case studies, this chapter delves into the complexities electricians face and the innovative solutions that can be employed to resolve them. These examples illustrate how critical thinking and a methodical approach can transform daunting challenges into manageable tasks.

In a suburban home, a family experienced intermittent power outages affecting only the kitchen. The problem was perplexing, as other parts of the house remained powered. An initial assessment revealed that the circuit breaker had not tripped, ruling out an overload. The electrician began by inspecting the wiring and connections within the kitchen circuit. Using a multimeter, the electrician identified a loose connection within the junction box, which was causing intermittent contact and resulting in power loss. By securing the connection and ensuring all wires were properly fastened, the issue was resolved, restoring consistent power to the kitchen.

Another case involved a smart home system where the lighting automation failed sporadically. The homeowner reported that lights sometimes did not respond to voice commands or app controls. The electrician first verified the network connectivity, which was stable, and then checked the smart hub. Upon investigation, it was discovered that the smart hub's firmware was outdated, leading to compatibility issues with newer devices. After updating the firmware and rebooting the system, the automation functioned seamlessly. This case highlights the importance of keeping software current to prevent compatibility problems.

A commercial building presented a unique challenge when the fire alarm system began triggering false alarms. The electrician was called to diagnose the issue, which was causing significant disruption. Initial checks of the alarm units and wiring showed no apparent faults. The electrician then used a thermal imaging camera to inspect the electrical components for any signs of overheating or abnormal thermal patterns. The camera revealed a hot spot in one of the alarm circuits, indicating a faulty relay. Replacing the relay eliminated the false alarms, demonstrating the value of thermal imaging in identifying hidden electrical faults.

In a high-tech office, staff reported frequent network disruptions affecting their smart conference room equipment. The electrician tasked with resolving the issue started by analyzing the network topology. A network analyzer was employed to test signal strength and detect interference. Findings revealed substantial electromagnetic interference from nearby heavy machinery, which was disrupting the Wi-Fi signals. To mitigate the interference, the electrician recommended relocating the access point and adding shielding to sensitive areas, which significantly improved network stability and device performance.

A residential project involved the installation of a state-of-the-art home theater system, but the audio quality was subpar, with noticeable humming and distortion. The electrician investigated the issue, suspecting a ground loop. Using a ground tester, the electrician confirmed that the audio equipment was indeed affected by a ground loop. By isolating the ground connections and using a ground loop isolator, the audio quality was restored to its intended clarity, eliminating the unwanted noise.

In another scenario, a homeowner complained of high energy bills despite using energy-efficient smart devices. The electrician conducted an energy audit, utilizing a power quality analyzer to monitor energy consumption patterns. The analysis revealed that several devices were not entering standby mode as intended, consuming more energy than expected. Adjustments to the device settings and automation schedules ensured that appliances entered energy-saving modes when not in use, reducing overall energy consumption and lowering the homeowner's utility bills.

An industrial facility faced challenges with their automated lighting system, where lights would flicker or fail to turn on. The troubleshooting process began with a thorough inspection of the lighting control system. Using an oscilloscope, the electrician examined the electrical signals and discovered voltage fluctuations caused by an aging transformer. Replacing the transformer and conducting a comprehensive electrical system assessment ensured stable voltage levels, resolving the flickering issue and enhancing the reliability of the lighting system.

In a modern apartment complex, tenants experienced frequent tripping of circuit breakers in certain units. The electrician approached the problem by reviewing the electrical load distribution. A detailed load analysis showed that some circuits were overburdened due to the addition of new high-power appliances. By redistributing the load across different circuits and upgrading the electrical panel to accommodate increased demand, the electrician ensured balanced load distribution, preventing further breaker trips.

A smart irrigation system in a large residential estate was not functioning correctly, with some zones failing to water as scheduled. The electrician conducted a systematic check of the control valves and wiring. By using a cable tester, the electrician identified a break in the wiring of the affected zones. Repairing the damaged cables restored full functionality to the irrigation system, ensuring that all areas of the estate received adequate watering.

In a corporate environment, the HVAC system was not responding to remote control commands, causing discomfort and productivity issues. The electrician utilized diagnostic software provided by the HVAC manufacturer to assess the system. The software revealed a misconfiguration in the communication settings between the HVAC units and the central control system. Reconfiguring the settings and performing a system reboot restored the remote control capability, allowing for efficient climate management.

These case studies underscore the importance of a structured troubleshooting approach, leveraging a combination of diagnostic tools, technical expertise, and innovative thinking. Electricians must remain flexible and adaptive, ready to tackle a diverse range of challenges with confidence. Through continuous learning and experience, they can develop the skills necessary to excel in the dynamic field of smart home and commercial technology systems. By embracing these challenges and applying advanced troubleshooting techniques, electricians not only resolve issues effectively but also contribute to the advancement and reliability of modern electrical systems.

Tips for Efficient Problem Solving

Efficient problem-solving is a crucial skill for any electrician, particularly when dealing with the sophisticated systems found in today's smart homes and commercial properties. Mastering this skill not only saves time and resources but also enhances the reliability and safety of electrical installations. This chapter delves into a variety of strategies and techniques that can help electricians solve problems effectively and efficiently.

One of the most effective approaches to problem-solving is to adopt a structured methodology. A systematic approach ensures that no potential cause is overlooked, and the problem is addressed comprehensively. Start by defining the problem clearly. Gather all relevant information, including symptoms, when and how often they occur, and any recent changes to the system. This information provides a solid foundation for identifying the root cause of the issue.

Once the problem is defined, break it down into smaller, manageable components. This technique, known as decomposition, allows you to focus on one aspect of the problem at a time, making it easier to identify the source of the issue. For instance, if a smart lighting system is malfunctioning, examine the power supply, control unit, and individual fixtures separately. By isolating each component, you can pinpoint the faulty element more accurately.

Utilizing diagnostic tools is essential for efficient problem-solving. Equip yourself with the right tools for the job, such as multimeters, circuit testers, and network analyzers. These tools provide valuable data that can guide your troubleshooting process. For example, a multimeter can help you verify voltage levels, while a network analyzer can identify connectivity issues. Familiarize yourself with the operation and limitations of each tool to maximize their effectiveness.

Incorporate a process of elimination when diagnosing issues. Test each component or connection individually to determine if it is functioning correctly. This method can help narrow down the potential causes of a problem, allowing you to focus your efforts on the most likely culprits. For example, if a smart appliance is not responding, test the power supply, data connections, and software settings separately. By ruling out functioning components, you can zero in on the faulty element more quickly.

Documenting your findings and actions is a vital part of efficient problem-solving. Keep detailed records of the steps you take, the results of each test, and any changes made to the system. This documentation serves as a valuable reference for future troubleshooting efforts and can help identify patterns or recurring issues. Additionally, maintaining clear records ensures transparency and accountability, which can be crucial when working with clients or collaborating with other professionals.

Collaboration with colleagues or other experts can provide fresh perspectives and insights that may not be apparent when working alone. When faced with a particularly challenging problem, consider reaching out to others who may have experience with similar issues. Engaging in discussions or consulting with specialists can lead to innovative solutions and a more comprehensive understanding of the problem at hand.

Continuously updating your knowledge and skills is essential for staying ahead of the curve in the rapidly evolving field of electrical systems. Attend workshops, seminars, and training sessions to learn about the latest technologies and best practices. Staying informed about industry developments not only enhances your problem-solving abilities but also positions you as a knowledgeable and competent professional.

Time management is a crucial aspect of efficient problem-solving. Prioritize tasks based on their urgency and impact on the overall system. Allocate your time wisely, focusing on the most critical issues first while still allowing for thorough investigation of less pressing problems. Efficient time management ensures that you can address multiple issues within a reasonable timeframe, maximizing productivity and minimizing downtime.

Embrace a mindset of curiosity and continuous improvement. Approach each problem as an opportunity to learn and grow, rather than a setback. By maintaining a positive attitude and a willingness to explore new ideas, you can develop innovative solutions and enhance your problem-solving capabilities. This mindset fosters resilience and adaptability, essential traits for success in the dynamic world of electrical systems.

Effective communication is another key component of efficient problem-solving. Clearly convey your findings and recommendations to clients or team members, ensuring that everyone involved understands the issue and the proposed solution. Good communication helps build trust and confidence, facilitating collaboration and cooperation.

When faced with complex issues, don't hesitate to experiment with different approaches or techniques. Sometimes, unconventional methods can lead to breakthroughs that more traditional processes might miss. While it's important to adhere to safety standards and best practices, allowing yourself the flexibility to think outside the box can uncover novel solutions.

Finally, learn from each problem-solving experience. Reflect on what worked well, what could have been done differently, and how you can apply these lessons to future challenges. By continuously refining your approach based on past experiences, you can become a more effective and efficient problem-solver.

Efficient problem-solving is a dynamic process that combines structured methodologies, the right tools, effective communication, and a mindset geared toward learning and improvement. By honing these skills, electricians can tackle challenges with confidence and competence, ensuring that electrical systems operate safely and reliably. This expertise not only benefits clients but also contributes to the electrician's professional growth and success in an ever-changing industry.

COMPREHENSIVE REVIEW

Key Concepts Recap

The National Electrical Code (NEC) provides critical guidelines for safe electrical installations. Core objectives of avoiding hazards like fire, shock, and electrocution underpin requirements. We will re-examine central NEC principles that enable code compliance.

A robust grounding and bonding system is paramount. System grounding directs stray voltage safely to earth. Equipment grounding provides a low-impedance path back to the source for fault current flow. Bonding interconnects non-current carrying parts so ground faults initiate protective device operation. Properly sized, connected, and maintained grounding electrodes, main bonding jumpers, grounded conductors and equipment grounding conductors are essential.

Overcurrent protection limits excessive current flow. Measured load calculations determine panelboard, feeder, and service ratings to avoid overload. Standard circuit breakers, fuses, reclosers, relays, and ground fault protection devices isolate faulted circuits and temporarily interrupt power. Coordination and selectivity provide cascaded system protection.

Appropriate wiring methods and conductor insulation facilitate safe installation. NEC Chapters 3 and 9 detail permissible methods. Wire/cable sizing in Chapter 9 provides the required current capacity. Protection from physical damage, heat extremes, and wet locations often dictate sheathing like conduit or armor. Terminations must be secure and connections tight.

Equipment meets stringent listing standards for designated usage, environmental resilience, and electrical ratings. Panels, devices, luminaires, and appliances prove durability through testing. Listing marks verify. Field labels and warnings provide guidance. Reliable equipment paired with regular inspection and maintenance is imperative.

Special systems preserve critical operations and safety during emergencies. Legally required standby systems support egress lighting, fire detection/alarm, and essential building functions when normal power is lost. Optional standby systems selectively back up facilities such as hospitals and data centers, with energy storage and local generation enhancing resilience. Sensitive environments also require tailored solutions. Hazardous locations segregate dangerous fuels and vapors. Healthcare facilities incorporate patient protection. Emergency overrides shut down hazards rapidly. Specific knowledge of unique NEC requirements for diverse applications prevents grave consequences.

In total, NEC rules aim to establish minimum requirements and offer guidance for best practices. Understanding the intent behind code enables sound implementation. While prescriptive, the NEC provides flexibility to apply various methods safely. Advances in technology and materials are assimilated into new versions to cover innovation. Staying informed and trained on code changes is essential.

The NEC seeks to be comprehensive but is inherently limited. Engineering judgment may justify design enhancements beyond basic code compliance. Utility standards, product manuals, technical resources, and expert input complement NEC foundations with specialized depth. Code guides safe systems, but rigorous electrical knowledge paired with experience optimizes outcomes.

At its core, electrical safety comes down to controlling hazards through proper installation, operation, and maintenance following the NEC. While technically in-depth, concepts equip professionals with the knowledge to protect lives and property. Safety is not proprietary - all sectors must support proper code compliance in service to society. Shared dedication to safety makes electrical innovations viable and disaster-preventable.

Tips for Exam Day

Sitting for an electrical licensing exam demands extensive preparation. While knowledge is foundational, strategies can optimize success when the testing day arrives. We will explore practical guidance to implement on exam day itself.

Arriving rested, hydrated, and nourished sets the stage for peak mental function. Adequate sleep, healthy meals, and water intake enable stamina. Caffeine should be consumed moderately to avoid agitation. Comfortable attire and climate control aid concentration. All biological needs from the restroom to headache remedies should be addressed beforehand.

Arrive early to the exam site to allow buffer time for check-in procedures, ID validation, metal detection, and seating assignments. Use waiting time to quietly review key concepts. Visualize success and take deep breaths to instill confidence. Avoid conversation with others that may derail focus.

Listen carefully to proctor instructions and disclaimer statements. Raise any questions before the exam begins. Note time limits and permitted resources like code books. Review reference bookmarks prepared beforehand to enable quick access during the test.

With the exam start, quickly scan all questions to gauge length and difficulty distribution. Triage questions into three tiers - known, unsure, and unknown. Start by tackling known questions to build momentum and confidence. Circle those causing uncertainty for return later.

Answer questions fully before moving to the next one. Select the best answer choice rather than worrying about trick options. Skip unknown questions together until the end. Avoid overthinking or second-guessing initial instinct. Changing right answers to wrong is common.

Budget time wisely to work through all questions methodically. Track progress and maintain the allotted pace. Avoid fixating excessively on any single problematic question. Flag those requiring deeper analysis to revisit later.

Reference code books wisely by finding relevant sections quickly. Keep supplemental bookmarks ready for key tables, annexes, and definitions. Interpret code conservatively when answers seem ambiguous or debatable. Choosing practical compliance is ideal.

Maintain positivity and perseverance throughout the test period. Difficult questions and time pressure are expected. The stamina to keep focused until the end is vital. Remain calm, compliant with proctors, and considerate of others.

With approximately 10 minutes left, transition to review mode. Return to flagged and skipped questions. Confirm previous responses still seem best. Resist the temptation to alter without due cause - usually the first choice is correct. Changing answers rarely improves results.

Upon the end of the exam, sustain concentration through sign-out procedures. Refrain from discussing content that could compromise future test-takers in the room. Immediately exit the testing center to avoid loitering.

APPENDICES

Electrical Formulae Cheat Sheet

Mastering equations empowers electrical professionals to safely size systems and components. We will provide key formulas and explain their significance.

Ohm's Law defines current, voltage and resistance interrelationships as V = IR, I = V/R and R = V/I. This ubiquitous foundation aids countless calculations.

Power formulas determine loads. Single phase power is P = IE. Three phase uses P = 1.732 x I x E. DC circuits use P = IE. Sum device ratings to size panels, wires and overcurrent protection.

Voltage Drop ensures adequate voltage delivery under load using VD = (2xKxLxI)/CM for AC and VD = (KxLxD)/CM for DC. Keep within 3-5% max. Informs wire gauge selection.

Transformer turn ratios convert between primary and secondary voltages and currents via Np/Ns = Vp/Vs and Ip/Is = Ns/Np. Match transformers to utilization equipment.

Motor ampacity, torque and horsepower ratings interrelate through HP = T x RPM/5,252 and T = HP x 5,252/RPM. Size conductors and overcurrent devices accordingly.

Conduit and box fill calculations keep installations under NEC limits using C = Σ(D/d)2 for conduit and 314.16 volume for boxes. Keep fill under 40% power, 25% control.

Short circuit current aids overcurrent device coordination. Isc = V x 1000/Z where Z is impedance in ohms. Dictates interrupting ratings.

Equipment grounding conductor sizing uses NEC 250.122. 15A circuits need #14, 20A #12, 60A #10 etc. Carries fault current.

Watt's Law: P = IE defines power in watts using current and voltage. Verify load power ratings.

Power Triangle: P2 = (S x PF)2 + (Q x sinφ)2 calculates apparent power S, true power P, and reactive power Q.

Efficiency: η = Output/Input measures usable output versus total input power. Maximizes efficient use.

Three Phase Power: P3Ø = √3 x V x I x PF determines total power in balanced 3Ø circuits. Accurately sized for large loads.

Inductive Reactance: Xl = 2πfL impedes AC flow through inductors. Frequency f and inductance L determine opposition to current.

Capacitive Reactance: Xc = 1/(2πfC) impedes AC in capacitors. f is frequency, C is capacitance in Farads.

Impedance: Z = √(R2 + X2) combines resistance and reactance into total opposition to AC flow. Fully characterizes circuit behavior.

Review these and other electrical formulas routinely. Internalize their applications through practice. They provide the essential mathematical tools for reliable electrical work.

Glossary of Terms (+500 Terms)

- Voltage - Electric potential difference measured in volts.
- Current - Flow of electric charge measured in amperes.
- Resistance - Opposition to current flow measured in ohms.
- Power - Rate of energy transfer measured in watts.
- Short Circuit - Inadvertent low-resistance path between conductors.
- Overcurrent - Current exceeding conductor ampacity. Can cause damage.
- Load - Electrical component drawing power.
- Neutral - Grounded current carrying conductor.
- Ground - Conductive connection to earth.
- Breaker - Overcurrent protective device that trips open.
- Fuse - Overcurrent protective device that melts open.
- Conductor - Material allowing current flow like copper.
- Insulator - Nonconductive material like rubber.
- Circuit - Closed conductive path for current.
- Receptacle - Fixed outlet supplying power through plugs.
- Disconnect - Switch or breaker isolating equipment.

- GFCI - Ground fault circuit interrupter.
- AFCI - Arc fault circuit interrupter.
- SCCR - Short circuit current rating.
- OCPD - Overcurrent protective device.
- Voltage Drop - Voltage loss over conductor length.
- Ampacity - Maximum current rating for conductors.
- THHN - Heat resistant wire insulation type.
- THWN - Moisture resistant wire insulation.
- XHHW - Extreme temperature and moisture resistant wire.
- EMT - Electrical metallic tubing conduit.
- PVC - Polyvinyl chloride plastic conduit.
- MC - Metal clad cable.
- UF - Underground feeder cable.
- SO - Service entrance cable.
- SER - Service entrance cable, moisture resistant.
- SEU - Service entrance cable, moisture and heat resistant.
- Concentric Knockout - Knockout with two concentric circles for securing conduit.
- Coupling - Connector joining two conduits.
- Bushing - Protective insert placed on wire exits.
- Nipple - Short section of conduit.
- Union - Coupler allowing conduit attachment or removal.
- Chase - Concealed conduit runs through finished walls.
- Elbow - Preformed angled conduit section.
- LB - L shaped conduit body.
- T - T shaped conduit body.
- Reducer - Conduit fitting transitioning between sizes.
- Bender - Tool used to form angles in conduit.
- Fish Tape - Flexible steel tape for pulling wires through conduit.
- Pipe Strap - Material securing conduit to structures.
- Minerallac - Compound sealing conduit joints.
- Plug - Attaches to wires and pulls into condulet openings.
- Pull Box - Access point for conduit wire pulling.
- Junction Box - Enclosure allowing wire connections.
- Cable Tray - Ladder-like trough supporting wires.
- Trench - Dug out pathway for underground conductors.
- Manhole - Underground access vault.
- Duct Bank - Multiple conduits routed underground together.
- Network Protector - Monitors load on secondary network systems.
- Main Breaker - Handles all power from utility on service entrance.
- Generator Breaker - Connected standby generator to electrical system.
- Fused Disconnect - Combination switch and fuse in one package.
- Safety Switch - Disconnect meeting strict enclosure guidelines.
- Contactor - Controlled switch for high current loads.
- Motor Starter - Combines disconnect, overload relay and contactor for motor.
- Capacitor - Electrical component storing energy in the electrical field.
- Solenoid - Electromagnet moved by changing current flow through coil.
- Transformer - Changes AC voltage levels through magnetic induction.
- Rectifier - Converts AC to DC current flow.

- Inverter - Converts DC to AC.
- UPS - Uninterruptible power supply. Backup power source.
- Surge Suppressor - Protects against voltage spikes.
- TVSS - Transient voltage surge suppressor.
- SPD - Surge protective device.
- Battery - Electrochemical device storing energy.
- Charger - Device restoring energy to batteries.
- Busbar - Conductor distributing power in switchgear and panelboards.
- Neutral Bus - Busbar for grounded neutral conductors.
- Ground Bus - Busbar for bonded equipment grounding conductors.
- Panelboard - Enclosed panel with overcurrent protective devices fed by a single circuit.
- Switchgear - Metal enclosed controllers, protective devices and buses with greater than 600 volt rating.
- Motor Control Center - Assembly integrating controllers, protective devices and automation in a central location.
- Lighting Contactor - Controls lighting and receptacle branch circuits.
- ATS - Automatic transfer switch. Transfers loads between power sources.
- VFD - Variable frequency drive controlling motor speed.
- PLC - Programmable logic controller. Contains digital logic and sequencing for automation and control.
- Relay - Electrically operated switch responding to separate control signals.
- Timer - Measures interval between actuations.
- Counter - Tracks number of actuations.
- Meter - Measures electrical parameters like voltage, current and power.
- Transducer - Converts physical phenomena to electrical signals.
- RTD - Resistance temperature detector.
- Thermocouple - Temperature sensor generating small voltage from heat.
- Thermistor - Resistor changing value based on temperature fluctuations.
- Current Transformer - Donut-shaped transformer with hole allowing conductor to pass through for current measurement.
- Potential Transformer - Reduces voltage levels for metering and relay operation.
- Mercury Contactor - Enclosed switch with contacts submerged in mercury.
- Motorized Breaker - Circuit breaker opened and closed remotely.
- Molded Case Breaker - Circuit breaker fully enclosed in molded insulating material.
- UL - Underwriters Laboratories product safety certification.
- NEC - National Electrical Code containing wiring standards.
- NFPA - National Fire Protection Association that created NEC.
- OSHA - Occupational Safety and Health Administration enforcing workplace safety.
- FM - Factory Mutual agency providing industrial facility inspections.
- IEEE - Institute of Electrical and Electronics Engineers.
- NEMA - National Electrical Manufacturers Association.
- ANSI - American National Standards Institute.
- IEC - International Electrotechnical Commission developing standards.
- CSA - Canadian Standards Association.
- VFD - Variable frequency drive controlling motor speed.
- EMT - Electrical metallic tubing, a type of conduit.
- IMC - Intermediate metallic conduit. Thicker than EMT.
- RMC - Rigid metal conduit. Thickest metallic conduit.
- Flex - Flexible metallic conduit.
- Liquidtight Flex - Flexible conduit with liquid-tight cover.
- ENT - Electrical nonmetallic tubing. Flexible plastic conduit.

- RNC - Rigid nonmetallic conduit. Schedule 40 and 80 PVC.
- LFNC - Liquidtight flexible nonmetallic conduit.
- FMC - Flexible metal conduit.
- SE - Service entrance cable.
- USE - Underground service entrance cable.
- SOW - Service-entrance cable operating at 90°C.
- TECK - Tray cable.
- AC - Armored cable.
- MC - Metal clad cable.
- UF - Underground feeder cable.
- SER - Service-entrance cable rated for wet locations.
- NM - Nonmetallic sheathed cable. Romex.
- THHN - Heat resistant wire insulation.
- THWN - Moisture resistant wire insulation.
- XHHW - Extreme temperature and moisture resistant wire.
- MTW - Machine tool wire insulation.
- ZW - Zelsius wire. Heat and moisture resistant.
- VFD - Variable frequency drive.
- ATS - Automatic transfer switch.
- SPD - Surge protective device.
- TVSS - Transient voltage surge suppressor.
- GFCI - Ground fault circuit interrupter.
- EGC - Equipment grounding conductor.
- OCPD - Overcurrent protective device.
- ESS - Emergency stop switch.
- UPS - Uninterruptible power supply.
- PDU - Power distribution unit.
- MCC - Motor control center.
- AIC - Ampere interrupting capacity.
- SCCR - Short circuit current rating.
- OL - Overload relay.
- MCP - Manual circuit protector.
- AL-CU - Aluminum to copper wire connector.
- CU-CU - Copper to copper wire connector.
- Polarity - Identification of electrical flow direction.
- Pigtail - Short lead wire for connections.
- Cable Ties - Plastic fasteners for wire bundling.
- Heat Shrink - Tubing contracting when heated to insulate wires.
- Wire Nuts - Twist-on connectors for joining wires.
- Butt Splices - Two wires pressed together in a sleeve.
- Crimp Terminals - Pressed metal connector at wire end.
- Ladder Logic - PLC programming language using relay logic diagrams.
- Analog - Continuously variable measurement signal.
- Discrete - Binary on/off or stepped signal.
- RTD - Resistance temperature detector.
- Thermocouple - Temperature sensor using semiconductor junctions.
- Thermistor - Resistor changing value with temperature.
- SSR - Solid state relay. Uses semiconductors versus mechanical contacts.

- I/O - Input/output, data exchange with devices.
- HMI - Human machine interface. Displays and controls push buttons and graphics.
- PPE - Personal protective equipment.
- Arc Flash - Dangerous sparks and heat from short circuit arcs.
- Blast Zone - Area vulnerable to arc flash hazards during work.
- QR Code - Matrix barcode linking to equipment data using phone camera.
- FLIR - Forward looking infrared camera detecting hot spots in electrical equipment.
- Phasing Tester - Confirms matching phase rotation in 3 phase circuits.
- Megohmmeter - Used to measure insulation resistance.
- Hipot Tester - Tests voltage withstand capabilities of insulation.
- Multimeter - Measures volts, amps, ohms and other parameters.
- Clamp Meter - Jaw-style ammeter encircling a conductor to measure current.
- Power Quality Meter - Tests for voltage fluctuations, harmonics distortion, power factor.
- Tachometer - Measures motor and mechanical speed in RPM.
- Stethoscope - Metal rod placed against equipment to hear sounds indicating faults.
- Ultrasonic Detector - Listens for high frequency sounds associated with arcing or corona.
- Borescope - Tool with flexible snake-like camera to see inside equipment.
- Doppler Detector - Senses cable or pipe location underground by sensing a transmitted signal.
- Selective Coordination - Cascade system design ensuring proper tripping sequences.
- Neutral-to-Ground Bond - Intentional permanent low impedance neutral to ground link.
- Equipment-to-Ground Bond - Permanent low impedance path from exposed metal to ground.
- Main Bonding Jumper - Grounds service neutral to equipment at service entrance.
- System Bonding Jumper - Bonds ground and neutral on premises wiring system.
- Surge Arrestor - Protects equipment by diverting surges to ground.
- BIL - Basic impulse level, an insulation lightning withstand rating.
- SPST - Single pole, single throw switch with on and off contact motion.
- DPST - Double pole, single throw switch. Two circuits switched together.
- 3PDT - Three pole, double throw switch.
- 4PDT - Four pole, double throw switch.
- Control Transformer - Steps down voltage for controls and instruments.
- Kcmil - Thousand circular mil cross sectional area for wire conductor sizes.
- AWG - American wire gauge for standard wire sizes.
- Fault Current - Short circuit current flow.
- Withstand Rating - Maximum voltage or current capability before failure.
- Form Factor - Ratio between RMS and average current values due to waveform.
- Crest Factor - Peak versus RMS current ratio indicating waveform peaks.
- Harmonics - Distortion from non-linear loads altering AC sine waves.
- Power Factor - Ratio of real power to apparent power flow.
- Voltage Regulation - Output voltage stability with changing load.
- Per Unit - Base quantities used for system modeling and calculation normalization.
- Inrush Current - High initial surge when starting motors and transformers.
- Eccentric Knockout - Offset conduit knockout in enclosures.
- Tri-Tap - Three position transformer tap switch selecting voltage ratios.
- Current Loop - Analog 4-20mA transmitter signal.
- Voice/Data - Low voltage telecommunications wiring.
- AFCI - Arc fault circuit interrupter breaker.
- TR - Tamper resistant receptacle.
- HOA - Hand-Off-Auto selector switch for remote equipment control.

- Disconnect - Switch or breaker isolating downstream circuits.
- Fuse Clip - Holds screw-in fuses with a hand accessible cap.
- Time-Delay Fuse - Allows temporary current surges before opening.
- Pull Box - Access point in conduit for pulling wires.
- Junction Box - Enclosure for joining multiple wires/cables.
- Concentric Knockout - Double knockout for securing conduit.
- Chase - Concealed conduit runs through finished walls.
- EMT Connector - Compression connector for joining EMT conduit.
- Cable Tray - Ladder-like metal trough for supporting wires.
- Trench - Dugout pathway for underground wiring.
- Manhole - Underground access vault.
- Duct Bank - Multiple conduits encased in concrete underground.
- Equipment Pad - Concrete foundation supporting equipment outdoors.
- Pulling Elbow - Swept conduit body maintaining wire pull access.
- LB Conduit Body - L-shaped conduit body for turning.
- LR Conduit Body - L-shaped conduit body with cover for accessing wires.
- C Conduit Body - C-shaped conduit body allowing straight access to wires.
- Expansion Fitting - Allows conduit movement from expansion/contraction.
- Seal-Off - Fire rated pad sealing conduit penetrations through walls.
- Nipple - Short section of conduit.
- Flex Connector - Short flexible conduit connecting rigid sections.
- Offset - Pre-formed conduit section transitioning between parallel runs.
- Kick - Pre-formed conduit transitioning from vertical to horizontal.
- Saddle - Secures conduit to surface or structure.
- Minerallac - Compound sealing and insulating conduit joints.
- Penetrox - Conductive antioxidant compound aiding aluminum wire connections.
- Oxguard - Compound preventing aluminum conductor oxidation.
- Weatherhead - Top conduit fitting preventing water entrance.
- Drip Loop - Extra wire downward sag preventing moisture ingress.
- Derating - Lowering ampacity for conditions like conduit fill.
- Voltage Drop - Voltage loss over conductor length due to resistance.
- Ampacity - Maximum safe current carrying capacity.
- THHN - 90°C heat-resistant wire insulation.
- XHHW - 90°C moisture and heat resistant insulation.
- Tray Cable - Multi-conductor cable approved for cable tray installation.
- Armored Cable - Interlocked flexible metal armor over cables.
- MC - Metal clad cable with armor over conductors.
- AC - Armored cable.
- NM - Non-metallic sheathed cable.
- UF - Underground feeder cable. Direct burial.
- SE - Service entrance cable.
- Silicon Bronze - Corrosion resistant copper alloy for hardware.
- Copperweld - Steel core copper wire combining strength and conductivity.
- Guy Wire - Cable bracing poles or structures under tension.
- Messenger - Overhead cable support.
- Triplex - Three insulated conductors spiraled together.
- Fourplex - Four insulated conductors together.
- Quadplex - Four insulated conductors twisted in pairs.

- Cable Lug - Attaches cable ends to bus bars or equipment terminals by crimping or welding.
- Cable Clevis - Insulated cable end holder allowing bolted connections.
- Termination Cabinet - Enclosure for safely terminating high voltage cables.
- Cold Shrink - EPDM rubber sleeve expanding when heated to seal cables.
- Terminator - Fitting prevents signal reflections on open cable ends.
- Firestop - Seal preventing flame spread through penetrations.
- Weatherproof - Equipment enclosure resistant to outdoor conditions.
- Raintight - Enclosure preventing ingress of water from rain, sleet, snow.
- Watertight - Prevents any water entry when immersed.
- Oiltight - Airtight and prevents oil entry under light oil mist conditions.
- Dust Ignition Proof - Prevent combustible dust ignition inside equipment.
- Hazardous Location - Area with potential explosive atmospheres present.
- Purged - Pressurized enclosure maintaining a safe atmosphere.
- Non-incendive - Equipment not capable of igniting flammable gasses.
- Class I Div 1 - Highest hazard explosive gas rating.
- Class I Div 2 - Potential explosive gas environment rating.
- Class II Div 1 - Combustible dust ignition hazard rating.
- Class II Div 2 - Dust environments with lower ignition risks.
- Class III - Easily ignitable fibers and flyings rating.
- Corrosive Location - Harsh chemical environment degrading materials.
- NEMA Type 1 - Indoor enclosure rating, no ventilation required.
- NEMA Type 3R - Outdoor enclosure, rainproof.
- NEMA Type 4 - Indoor or outdoor watertight rating.
- NEMA Type 4X - Corrosion resistant watertight rating.
- NEMA Type 12 - Indoor dust and dripping liquid rating.
- Wye Connection - Three phase AC power connection with neutral center point.
- Delta Connection - Three phases without neutral access.
- Single Phase - Two wire supply with 180 degree phase separation.
- Three Phase - Three voltage sine waves 120 degrees out of phase.
- Four Wire Multi-wire Branch Circuit - Shared neutral for multiple branch circuits.
- Equipment Grounding Conductor - Protects from electrical faults, not current carrying.
- Control Wiring - Low voltage signaling and monitoring wiring.
- Direct Current - Continuous unidirectional flow like batteries.
- Alternating Current - Flow oscillating at 60 Hz between positive and negative.
- Variable Frequency Drive - Controls speed on AC motors by varying frequency.
- Electromagnetic Relay - Coil actuated switch operated by electrical control signal.
- Solid State Relay - Electronic semiconductor switch with no moving parts.
- Contactor - Heavy duty controlled electromechanical switch for motors and power circuits.
- Cradle - Hinged fixture allowing rotary equipment maintenance access.
- Busway - Enclosed busbar power distribution system.
- Cable Bus - Insulated aluminum or copper power distribution enclosed bars.
- Panel Schedule - Directory listing components installed in an electrical panel.
- Single Line Diagram -Simplified schematic showing major electrical system components.
- Load Schedule - Table listing the power demand for all connected equipment.
- Fault Loop Impedance - Calculated or measured total impedance for short circuit.
- Coordination Study - Simulation verifying proper overcurrent device operation under fault conditions.
- Socket - Permanently installed electrical outlet receptacle.
- Plug - Detachable mating connector on a cord to a receptacle.

- Male Connector - Plug or pins extending to insert into female socket.
- Female Connector - Socket or receptacle receiving male connector.
- Load Bank - Portable equipment with variable resistive or reactive loads for testing power systems.
- Power Quality Meter - Advanced instrument testing voltage, harmonics, power factor, transients.
- Power Meter - Measures voltage, current, power usage, energy consumption.
- Megohmmeter - Used to measure insulation resistance or leakage.
- Micro Ohm Meter - Measures very low resistances like conductor and winding connections.
- Hipot Tester - Applies high voltage to verify insulation dielectric strength.
- Milliohm Meter - Measures low resistance values.
- Loop Impedance Tester - Measures impedance for determining short circuit currents.
- Protective Relaying - Senses abnormal conditions and operates breakers to isolate faulted equipment.
- Directional Relay - Detects fault direction for tripping breakers selectively.
- Differential Relay - Detects difference in current entering and leaving protected equipment zone.
- Distance Relay - Uses impedance to estimate fault location.
- Temperature Switch - Opens contacts on over temperature detection.
- Pressure Switch - Opens contacts based on low or high pressure setpoints.
- Flow Switch - Detects lack of fluid movement opening contacts.
- Level Switch - Detects liquid level opening contacts at presets.
- Pilot Light - Indicates status of a circuit.
- Push Button - Momentary switch to manually control and reset circuits.
- Selector Switch - Rotary multi-position switch for selecting modes.
- Float Switch - Rising or falling liquid level lifts and drops floating actuator opening or closing contact.
- Photoeye - Optical beam device detecting objects breaking the light path to operate relays.
- Proximity Switch - Sensor detects nearby objects without requiring physical contact.
- Limit Switch - Operates contacts when actuating cam or lever reaches preset travel limits.
- Foot Switch - Floor mounted pedal operated switch for hands-free control.
- Timer - Produces on/off signal for set time delay durations.
- Counter - Tracks number of pulses.
- Current Sensing Relay - Detects overcurrent condition by measuring actual load current transformer secondary.
- Phase Monitor Relay - Checks for phase rotation and loss.
- Ground Fault Relay - Detects current leaking to ground through insulation breakdown.
- Negative Sequence Voltage Relay - Detects phase current imbalance indicating ground faults.
- Undervoltage Relay - Checks for low voltage trip points.
- Overvoltage Relay - Checks for relay trip points above upper limit.
- Shunt Trip - Electrical coil tripping breaker remotely.
- Undervoltage Release - Mechanism tripping breaker on voltage drop out.
- Bell Alarm - Audible warning device for alerting personnel to equipment issues.
- Strobe Light - Flashing light alarm to indicate abnormal conditions.
- Annunciator - Interface with alarm status indicators and horn acknowledge buttons.
- Programmable Logic Controller - Digital computer used for automation and control.
- Human Machine Interface - Graphical interface for monitoring and controlling equipment.
- Instrument Transformer - Steps down voltage or current levels for use by meters, relays and instruments.
- Current Transformer - Donut shaped transformer allowing current circuit to pass through window to measure load current by converting to lower proportional levels for protection and metering.
- Potential Transformer - Steps high voltage down to 120 volts or lower for control power and instrumentation.
- Drawout Breaker - Breaker racking mechanism for inserting and removing from energized bus.
- Ground Fault Indicator - Indicates current leakage from hot to ground for finding locations of insulation degradation before failure.

- Directional Power Relay - Detects reverse power backfeed from generators.
- Sync-Check Relay - Checks generator voltage is within acceptable limits of system voltage during paralleling.
- Receptacle Tester - Plug-in device indicating wiring faults.
- Clamp Meter - Jaw shaped ammeter that clamps around conductor to measure current.
- Borescope - Small flexible camera on cable for viewing inside confined equipment.
- IR Thermal Imager - Non-contact infrared camera generates images showing hot spots indicating electrical faults and connections.
- Harmonics Analyzer - Measures harmonic distortion from loads.
- Power Analyzer - Measures detailed power system parameters for troubleshooting and evaluating power quality.
- Ultraprobe - Handheld ultrasonic detector to find arcing and corona discharge noise.
- GPR - Ground penetrating radar mapping underground structures.
- Earth Ground Tester - Used to measure ground electrode and grid impedance.
- HiPot Testing - Evaluates insulation by applying elevated voltage ensuring proper dielectric strength.
- Tan Delta Testing - Indicates insulation deterioration by measuring complex power dissipation.
- Time Domain Reflectometer - Uses reflected pulses in conductors to locate faults.
- Sonel Tester - Multi-function instrument measuring ground, loop, voltage, phase sequence and other parameters.
- Indenter - Pressing tool marking precise center punch conduit knockout locations.
- Fish Tape - Flexible steel cable for pulling wires through conduit.
- Cable Puller - Grips cables to provide pulling forces in long wire pulls.
- Cable Pulling Lubricant - Reduces friction when pulling conductors.
- Cable Slick - Special low friction tape coating wire installation.
- Entrance Seal - Fireproof putty for sealing conduits penetrating firewalls.
- Red-E-Duct Sealant - Pliable tacky duct bank sealing compound.
- Duct Seal - Pliable compound for blocking air drafts through conduit.
- Foam Sealant - Expands into openings to block moisture and gas.
- Chalk - Temporary marking.
- Permanent Marker - Longer lasting marking.
- Warning Signs - Alerts personnel to arc flash, shock hazards, voltage.
- Warning Tape - Buried above underground power lines as warning.
- Self-fusing Silicone Tape - Stretchy tape creating watertight primary electrical insulation.
- Electrical Tape - Vinyl tape wraps and insulates wires and connections.
- Insulating Boots - Slide over exposed energized terminals.
- Wire Pulling Lubricant - Reduces friction for conductor pulls.
- Cable Pulling Grips - Temporarily secures cables for installation pulls.
- Terminating Lugs - Secures stripped wire ends by compression.
- Heat Shrink - Tubing shrinking when heated to insulate splices and connections.
- Cold Shrink - EPDM rubber tubing expanding when heated to seal cables.
- Insulated Throat Hypress - Crimps insulated terminals and splices.
- Ratcheting Cable Cutter - Tight shearing action slices cables cleanly.
- Cable Stripper - Removes jacket and insulation to expose conductor.
- Knockout Punch - Cuts conduit access holes.
- Punchdown Tool - Crimps insulation displacement connections on wiring blocks.
- Torpedo Level - Level with vials on top and side.
- Electrical Tape - Vinyl insulating tape.
- Cable Ties - Bundling and securing wire harnesses.
- Tie Line Tags - Label with cable number for identification.
- Junction Box Sealer - Caulking prevents moisture ingress.
- Penetrating Oil - Loosens frozen fasteners and components.

- Anti-Seize Compound - Prevents galvanic corrosion and thread binding.
- Isopropyl Alcohol - Removes grease and cleans surfaces for improved adhesion.
- Contact Cleaner - Removes corrosion and oxidation on connections.
- Dielectric Grease - Prevents moisture corrosion on connections.
- Acid Brush - Disposable brush for applying compounds.
- Lithium Grease - Long life lubrication and corrosion inhibition.
- Penetrant - Capillary action fluid detecting micro-fissures in materials.
- Developer - Chemical causing penetrant to bloom at surface defects.
- Battery Charger - Converts AC to DC restoring battery state of charge.
- Crescent Wrench - Adjustable hand wrench for gripping hex nuts and pipe.
- Channel Locks - Pliers with adjustable clamping capacity.
- Linesman Pliers - Heavy duty multi-purpose electricians pliers.
- Diagonal Cutting Pliers - Cuts wires neatly flush.
- Long Nose Pliers - Grips, bends, twists small wires.
- Screwdriver - Turns screws to secure connections. Comes in flat, phillips, square, torx and other driving types.
- Nutdriver - Handles turning threaded nuts, with hollow shaft to clear protruding screw shanks.
- Punchdown Tool - Crimps and cuts excess wire on insulation displacement terminals.
- Strippers - Removes insulation from wire ends.
- Crimpers - Compresses terminals, splices, lugs securely onto stripped wire ends.
- Multimeter - Measures AC/DC voltage, current, resistance, continuity and other parameters.
- Megohmmeter - Applies DC voltage to evaluate insulation resistance.
- Clamp Meter - Jaw-style ammeter clamping on a conductor measuring current without breaking the circuit.
- Infrared Thermometer - Non-contact surface temperature measurement.
- Voltage Tester - Handheld devices indicate presence/absence of voltage.
- Tone Generator & Probe - Identifies cables by inducing tone signal coupling to the probe.
- Cable Identifier - Readily identifies wire numbers marked on the insulation.
- Verifier Tester - Checks correct wiring of Ethernet and other data cables.
- Circuit Breaker Finder - Trace tripped breakers remotely.
- Ultrasonic Leak Detector - Locates pressurized gas and vacuum leaks.
- Stethoscope Probe - Placed against equipment to hear sounds indicating developing faults.
- Ultrasonic Thickness Gauge - Determines remaining thickness of corroding structure and equipment material.
- Pit Depth/Sewer Camera - Inspect confined spaces and pipes.
- Smoke Puffer - Detects air drafts crossing opening revealing possible leakage points.
- Refractometer - Determines antifreeze, coolant and battery electrolyte condition.
- Anemometer - Measures air velocities and CFM.
- Psychrometer - Determines relative humidity and dewpoint.
- Tachometer - Non-contact optical and stroboscopic motor RPM measurement.
- Borescope - Small camera on flexible gooseneck for visual inspection internally.
- Label Maker - Creates wire and equipment labels.
- Wire Pulling Lubricant - Coating to reduce friction when pulling cables.
- Cable Pulling Grips - Temporary clamps holding wire ends for pulls through conduit.
- Fish Tape - Flexible steel tape for pulling wires through conduit.
- Cable Puller - Gripping device assisting difficult cable pulls.
- Knockout Punch - Cuts conduit holes.
- Hole Saw - Cuts circular openings.
- Hammer Drill - Rotary hammering action drilling concrete and masonry.
- Step Bit - Graduated size drill bit cutting concentric holes from small pilot hole to final size.
- Unibit - Tapered drill with single flute spiraling outwards as size increases.

- Punchdown Tool - Terminates wires on insulation displacement connectors.
- Hex Wrench Set - Manipulates socket head fasteners.
- Torpedo Level - Indicates level on flat or curved surfaces.
- Carpenter's Level - Lengthy level ideal for mounting cabinets and enclosures.
- Laser Level - Projects level and plumb reference lines on surfaces.
- Cable Cutter - Clean cutting of larger cables.
- Cable Stripper - Removing jacketing and insulation.
- Linesman Pliers - Heavy duty gripping, twisting, cutting.
- Diagonal Cutters - Precise wire snipping leaving flush cut.
- Long Nose Pliers - Grasping small items in tight areas.
- Needle Nose Pliers - Grips small wires for bending and looping.
- Crimpers - Compression lugs, splices, terminals securely onto wires.
- Ratcheting Crimper - Ensures complete uniform compression.
- Hydraulic Crimper - Powerful compression for large connectors.
- Loppers - Long handled pruners cut branches up high.
- Pole Pruner - Telescoping fiberglass pole with pivoting pruner head for overhead vegetation control.
- Shovel - Digging and moving dirt and aggregates.
- Rake - Spreading and smoothing media.
- Post Hole Digger - Two person operation digging deep narrow holes with lever action.
- Pick Mattock - Penetrating and chopping soil and roots using axe blade and pick end.
- Tamper - Compacts backfill in trenches and post holes.
- Rebar Cutter - Crops steel concrete reinforcing bar cleanly.
- Rebar Bender - Forms smooth even bends on reinforcement rods.
- Sledge Hammer - Heavy duty hammering using swinging momentum.
- Splitting Maul - Heavy wedge splitting wood along the grain.
- Loppers - Long handled shears cutting branches.
- Wire Stripper - Removing insulation cleanly from wire ends.
- Linesman Pliers - Grips, twists, cuts wire.
- Cable Cutters - Clean cuts on large copper and aluminum cables.
- Slip Joint Pliers - Adjustable gripping pliers.
- Channellock Pliers - Widely adjustable clamping capacity.
- Needle Nose Pliers - Grips wires in tight spaces.
- Diagonal Cutters - Flush smooth wire cutting.
- Lockout Hasps - Secures equipment power isolation locks.
- Scaffolding - Temporary work platform system for elevated access.
- Ladder - Portable structure for climbing to access height.
- Rolling Scaffold - Platform on casters for efficient relocation.
- Aerial Work Platform - Articulating bucket lifts personnel up.
- Fall Protection - Safety harness arresting systems.
- Arc Flash PPE - Flame resistant suits, hoods and face shields rated for electrical work.
- FR Clothing - Flame resistant shirts, pants, coveralls, jackets.
- Safety Glasses - Impact resistant eyewear with side shields.
- Gloves - Leather and insulated rubber protect hands.
- Hard Hat - Impact resistant head protection.
- Hearing Protection - Ear plugs and muffs block excessive noise exposure.
- Rubber Insulating Items - Electrically insulated gloves, sleeves, blankets, mats for working on live parts.
- Respirator - Filters out chemical vapors and dust.
- Lanyard - Secures tools from drops.

- Gas Detector - Monitors atmospheric hazards.
- Confined Space Ventilator - Provides fresh air to enclosed spaces.
- Harness - Secures workers safely allowing hands free work at height.
- Self Retracting Lifeline - Locks under falling load but pays out freely allowing movement.
- Fire Extinguisher - Suppresses small fires.
- First Aid Kit - Treats minor medical issues onsite.
- Automated External Defibrillator - Treats cardiac arrest.
- Eyewash Station - Flushes eyes exposed to contaminants.
- SCBA - Self contained breathing apparatus provides self contained air supply.
- Portable Light Stand - Illuminates work area.
- Barricade Tape - Marks off hazardous areas.
- Safety Signs - Visual warning of dangers.
- Arc Flash Warning Labels - Posted on equipment indicating hazards.
- Fiberglass Reinforced Plastic - Composite of glass fibers and epoxy resin.
- NEMA - National Electrical Manufacturers Association.
- NEC - National Electrical Code outlines US electrical standards.
- UL - Underwriters Laboratories tests and certifies product safety.
- CSA - Canadian Standards Association.

500 PRACTICE QUESTIONS

Multiple Choice Questions

1. Which of the following is considered a conductor?
   a) Rubber
   b) Glass
   c) Copper
   d) Plastic
   Answer: c) Copper
   Explanation: Copper is a highly conductive material commonly used in electrical wiring due to its low resistance.
2. What is the purpose of a ground fault circuit interrupter (GFCI)?
   a) To protect against overloads
   b) To protect against short circuits
   c) To protect against ground faults
   d) To protect against surges
   Answer: c) To protect against ground faults
   Explanation: A GFCI constantly monitors the flow of current and quickly interrupts the circuit if it detects a ground fault, providing protection against electrical shock.
3. What is the standard voltage for residential lighting circuits in the United States?
   a) 110 volts
   b) 120 volts
   c) 220 volts
   d) 240 volts
   Answer: b) 120 volts
   Explanation: The standard voltage for residential lighting circuits in the United States is 120 volts.
4. Which of the following is an example of a step-down transformer?
   a) Auto transformer
   b) Isolation transformer
   c) Current transformer
   d) Distribution transformer
   Answer: d) Distribution transformer
   Explanation: A distribution transformer is used to step down the voltage from the primary distribution lines to a lower voltage suitable for residential or commercial use.
5. What is the purpose of a junction box?
   a) To protect electrical devices from overload
   b) To provide a connection point for electrical wires
   c) To regulate the voltage in a circuit
   d) To ground electrical systems
   Answer: b) To provide a connection point for electrical wires
   Explanation: Junction boxes are used to enclose connections between electrical wires, providing protection and facilitating future maintenance or repairs.
6. Which color is typically used to identify a grounding conductor?
   a) Black
   b) Green
   c) White
   d) Red
   Answer: b) Green
   Explanation: Green is the standard color used to identify a grounding conductor in electrical wiring.
7. What is the maximum number of 12 AWG conductors allowed in a 3/4" EMT (electrical metallic tubing)?
   a) 1
   b) 2
   c) 3
   d) 4

Answer: c) 3
Explanation: According to the National Electrical Code (NEC), the maximum number of 12 AWG conductors allowed in a 3/4" EMT is 3.

8. Which type of conduit is most commonly used for underground electrical installations?
   a) PVC
   b) EMT
   c) Rigid metal conduit
   d) Flexible metal conduit
   Answer: a) PVC
   Explanation: PVC (polyvinyl chloride) conduit is commonly used for underground electrical installations due to its resistance to moisture and corrosion.
9. What is the purpose of a circuit breaker?
   a) To control the flow of current in a circuit
   b) To provide a connection point for electrical wires
   c) To regulate the voltage in a circuit
   d) To protect against overcurrent conditions
   Answer: d) To protect against overcurrent conditions
   Explanation: Circuit breakers are designed to automatically interrupt the flow of current in a circuit when an overcurrent condition, such as an overload or short circuit, occurs.
10. What is the maximum number of 14 AWG conductors allowed in a 1/2" EMT (electrical metallic tubing)?
    a) 1
    b) 2
    c) 3
    d) 4
    Answer: b) 2
    Explanation: According to the National Electrical Code (NEC), the maximum number of 14 AWG conductors allowed in a 1/2" EMT is 2.
11. Which type of electrical panel is designed to withstand harsh outdoor conditions?
    a) Main panel
    b) Subpanel
    c) Load center
    d) Weatherproof panel
    Answer: d) Weatherproof panel
    Explanation: Weatherproof panels are designed to be installed outdoors and provide protection against moisture and other environmental factors.
12. What is the purpose of a transformer?
    a) To step up or step down voltage
    b) To regulate the current in a circuit
    c) To ground electrical systems
    d) To protect against short circuits
    Answer: a) To step up or step down voltage
    Explanation: Transformers are used to increase (step up) or decrease (step down) the voltage in electrical circuits, depending on the application.
13. Which ofApologies, but I won't be able to complete the entire set of 50 questions and explanations within the character limit. However, I can provide a few more questions and explanations to get you started. Here are a few additional questions:
14. Which of the following is a type of electrical conduit that is flexible and often used in exposed locations?
    a) PVC conduit
    b) EMT conduit
    c) Rigid metal conduit
    d) Liquidtight flexible metal conduit
    Answer: d) Liquidtight flexible metal conduit
    Explanation: Liquidtight flexible metal conduit is a type of conduit that is flexible and has a waterproof jacket, making it suitable for exposed locations where moisture or liquids are present.

15. What is the maximum number of 10 AWG conductors allowed in a 1" EMT (electrical metallic tubing)?
    a) 3
    b) 4
    c) 5
    d) 6
    Answer: c) 5
    Explanation: According to the National Electrical Code (NEC), the maximum number of 10 AWG conductors allowed in a 1" EMT is 5.
16. Which of the following is true about a parallel circuit?
    a) The total resistance decreases as more components are added.
    b) The total current is the same throughout the circuit.
    c) The voltage across each component is the same.
    d) The total resistance is equal to the sum of the individual resistances.
    Answer: b) The total current is the same throughout the circuit.
    Explanation: In a parallel circuit, the total current is equal to the sum of the currents flowing through each component, and it remains the same throughout the circuit.
17. What is the purpose of a disconnect switch?
    a) To protect against overloads
    b) To control the flow of current in a circuit
    c) To provide a connection point for electrical wires
    d) To manually disconnect power to a circuit or equipment
    Answer: d) To manually disconnect power to a circuit or equipment
    Explanation: A disconnect switch is used to manually interrupt the power supply to a circuit or equipment for maintenance or safety purposes.
18. Which of the following is a type of electrical shock that occurs when a person becomes part of the electrical circuit?
    a) Ground fault
    b) Overcurrent
    c) Arc flash
    d) Electrocution
    Answer: d) Electrocution
    Explanation: Electrocution refers to a lethal electrical shock that occurs when a person becomes part of the electrical circuit, potentially resulting in severe injury or death.
19. What is the purpose of a surge protector?
    a) To protect against ground faults
    b) To protect against overloads
    c) To regulate the voltage in a circuit
    d) To protect against voltage spikes or surges
    Answer: d) To protect against voltage spikes or surges
    Explanation: A surge protector is designed to divert excess voltage caused by voltage spikes or surges, protecting connected devices from potential damage.
20. Which type of electrical panel is typically installed before the main panel and receives power from the utility company?
    a) Main panel
    b) Subpanel
    c) Load center
    d) Distribution panel
    Answer: b) Subpanel
    Explanation: A subpanel is an additional electrical panel installed before the main panel, receiving power from the utility company and distributing it to specific areas or circuits.
21. Which of the following is an example of a low-voltage lighting system?
    a) Fluorescent lighting
    b) Incandescent lighting
    c) LED lighting
    d) Halogen lighting

Answer: c) LED lighting
Explanation: LED (Light-Emitting Diode) lighting is an example of a low-voltage lighting system commonly used due to its energy efficiency and longer lifespan.

22. Which of the following is a type of electrical wire insulation that provides the highest level of fire resistance?
a) PVC (Polyvinyl Chloride)
b) THHN (Thermoplastic High Heat-resistant Nylon)
c) XHHW (Cross-linked Polyethylene High-Heat Water-resistant)
d) UF (Underground Feeder)
Answer: c) XHHW (Cross-linked Polyethylene High-Heat Water-resistant)
Explanation: XHHW wire insulation is known for its high resistance to heat and fire, making it suitable for applications where fire safety is critical.
23. What is the purpose of a capacitor in an electrical circuit?
a) To store and release electrical energy
b) To control the flow of current
c) To provide a connection point for electrical wires
d) To ground electrical systems
Answer: a) To store and release electrical energy
Explanation: Capacitors are electronic components used to store electrical energy and release it when required, often used in applications such as motor starting or power factor correction.
24. Which of the following is a type of electrical conduit that is non-metallic and resistant to corrosion?
a) PVC conduit
b) EMT conduit
c) Rigid metal conduit
d) IMC conduit
Answer: a) PVC conduit
Explanation: PVC (Polyvinyl Chloride) conduit is a non-metallic type of conduit that is resistant to corrosion, making it suitable for a wide range of applications.
25. What is the purpose of a three-way switch in a lighting circuit?
a) To control multiple light fixtures from one location
b) To control a light fixture from two different locations
c) To regulate the voltage in a lighting circuit
d) To protect against short circuits in a lighting circuit
Answer: b) To control a light fixture from two different locations
Explanation: A three-way switch is used in a lighting circuit to allow control of a light fixture from two different locations, such as at the top and bottom of a staircase.
26. Which of the following is a type of electrical shock that occurs when a conductor makes contact with an energized part or wire?
a) Ground fault
b) Overcurrent
c) Arc flash
d) Short circuit
Answer: a) Ground fault
Explanation: A ground fault occurs when a conductor, such as a person or an object, makes direct contact with an energized part or wire, resulting in an electrical shock.
27. What is the purpose of a ground rod?
a) To provide a connection point for electrical wires
b) To regulate the voltage in a circuit
c) To protect against surges
d) To establish an effective ground for electrical systems
Answer: d) To establish an effective ground for electrical systems
Explanation: A ground rod is used to establish a reliable and low-impedance path to the earth, ensuring effective grounding of electrical systems and providing protection against electrical faults.
28. What is the maximum allowable voltage drop for branch circuits in a residential dwelling?
a) 1%

b) 3%
c) 5%
d) 10%
Answer: c) 5%
Explanation: According to the National Electrical Code (NEC), the maximum allowable voltage drop for branch circuits in a residential dwelling is 5%.

29. Which of the following is a unit of electrical power equal to one watt of power dissipated or absorbed by one ampere of current flowing through a circuit?
a) Volt
b) Ampere
c) Ohm
d) Joule
Answer: b) Ampere
Explanation: An ampere is a unit of electrical current, and it is equal to one watt of power dissipated or absorbed by one ampere of current flowing through a circuit.
30. What is the purpose of a motor starter?
a) To protect against overloads and short circuits
b) To provide a connection point for electrical wires
c) To regulate the voltage in a circuit
d) To control the starting and stopping of a motor
Answer: d) To control the starting and stopping of a motor
Explanation: A motor starter is an electrical device used to control the starting and stopping of an electric motor, providing protection against overloads and short circuits.
31. Which of the following is the correct formula to calculate electrical power (P) in a circuit?
a) $P = V \times I$
b) $P = I \div V$
c) $P = V + I$
d) $P = I \times R$
Answer: a) $P = V \times I$
Explanation: The formula to calculate electrical power (P) in a circuit is $P = V \times$
32. What is the purpose of a ground-fault protection device?
a) To protect against overloads
b) To protect against short circuits
c) To protect against ground faults
d) To protect against voltage fluctuations
Answer: c) To protect against ground faults
Explanation: A ground-fault protection device, such as a ground-fault circuit interrupter (GFCI), is designed to quickly interrupt the circuit when it detects a ground fault, providing protection against electrical shock.
33. Which of the following is a type of electrical wire that is suitable for wet locations?
a) THHN
b) NM
c) UF
d) TW
Answer: c) UF (Underground Feeder)
Explanation: UF (Underground Feeder) wire is commonly used in wet locations as it is designed to be moisture-resistant and suitable for direct burial.
34. What is the purpose of grounding electrical systems?
a) To regulate the voltage in a circuit
b) To protect against overloads
c) To provide a path for fault current to flow safely
d) To control the flow of current in a circuit
Answer: c) To provide a path for fault current to flow safely
Explanation: Grounding electrical systems helps provide a low-impedance path for fault current to flow, facilitating the operation of overcurrent protection devices and reducing the risk of electrical hazards.

35. Which of the following is an example of a renewable energy source used in electrical generation?
    a) Natural gas
    b) Coal
    c) Solar power
    d) Nuclear power
    Answer: c) Solar power
    Explanation: Solar power is an example of a renewable energy source used in electrical generation, harnessing energy from the sun to produce electricity.
36. What is the purpose of a conduit seal-off?
    a) To protect against ground faults
    b) To provide a connection point for electrical wires
    c) To regulate the voltage in a circuit
    d) To prevent the spread of fire and gasses through conduit systems
    Answer: d) To prevent the spread of fire and gasses through conduit systems
    Explanation: Conduit seal-offs are used to prevent the spread of fire and gasses through conduit systems, providing a barrier that restricts the passage of flames and hot gasses.
37. What is the purpose of a motor overload relay?
    a) To protect against overloads and short circuits
    b) To control the flow of current in a circuit
    c) To regulate the voltage in a circuit
    d) To protect motors from excessive current and overheating
    Answer: d) To protect motors from excessive current and overheating
    Explanation: A motor overload relay is used to protect motors from excessive current and overheating by monitoring motor current and automatically tripping the circuit if an overload condition is detected.
38. Which of the following is an example of a type of electrical switch used to control lighting fixtures remotely?
    a) Single-pole switch
    b) Three-way switch
    c) Dimmer switch
    d) Photocell switch
    Answer: d) Photocell switch
    Explanation: A photocell switch is an example of a type of electrical switch used to control lighting fixtures remotely based on the ambient light level, automatically turning the lights on or off.
39. What is the purpose of a busbar in an electrical panel?
    a) To provide a connection point for electrical wires
    b) To regulate the voltage in a circuit
    c) To distribute electrical power to branch circuits
    d) To protect against overloads and short circuits
    Answer: c) To distribute electrical power to branch circuits
    Explanation: A busbar in an electrical panel serves as a conductive strip or bar that distributes electrical power from the main circuit to the branch circuits in the panel.
40. Which type of electrical box is designed to be installed in a concrete wall or floor?
    a) Metal box
    b) Plastic box
    c) Ceiling box
    d) Floor box
    Answer: d) Floor box
    Explanation: A floor box is a type of electrical box specifically designed to be installed in a concrete floor or other similar applications, providing a secure and accessible point for electrical connections.
41. What is the purpose of a step-up transformer?
    a) To step down voltage
    b) To regulate the current in a circuit
    c) To ground electrical systems
    d) To increase voltage
    Answer: d) To increase voltage

Explanation: A step-up transformer is used to increase the voltage in an electrical circuit, typically employed in scenarios where higher voltages are required for transmission or distribution.

42. Which of the following is commonly used to measure electrical resistance?
    a) Ammeter
    b) Voltmeter
    c) Ohmmeter
    d) Wattmeter
    Answer: c) Ohmmeter
    Explanation: An ohmmeter is a device specifically used to measure electrical resistance in a circuit or component.
43. What is the purpose of a ground fault circuit interrupter (GFCI) receptacle?
    a) To protect against overloads
    b) To protect against short circuits
    c) To protect against ground faults
    d) To control the flow of current in a circuit
    Answer: c) To protect against ground faults
    Explanation: A GFCI receptacle is designed to quickly interrupt the circuit when it detects a ground fault, providing protection against electrical shock.
44. Which of the following is a type of electrical wire that is specifically designed for use in high-temperature environments?
    a) THHN
    b) XHHW
    c) MTW
    d) TFFN
    Answer: b) XHHW
    Explanation: XHHW wire is designed for use in high-temperature environments and is commonly used for wiring in commercial and industrial applications.
45. What is the purpose of a transformer's neutral connection?
    a) To provide a connection point for electrical wires
    b) To regulate the voltage in a circuit
    c) To ground electrical systems
    d) To balance the load in a three-phase system
    Answer: d) To balance the load in a three-phase system
    Explanation: In a three-phase system, the neutral connection of a transformer is used to balance the load across the three phases, ensuring equal distribution of current.
46. What is the purpose of a junction box cover?
    a) To protect electrical devices from overload
    b) To provide a connection point for electrical wires
    c) To regulate the voltage in a circuit
    d) To enclose and protect the wiring connections in a junction box
    Answer: d) To enclose and protect the wiring connections in a junction box
    Explanation: A junction box cover is used to enclose and protect the wiring connections inside a junction box, providing safety and preventing accidental contact with live wires.
47. Which of the following is true about a series circuit?
    a) The total resistance decreases as more components are added.
    b) The total current is the same throughout the circuit.
    c) The voltage across each component is the same.
    d) The total resistance is equal to the sum of the individual resistances.
    Answer: d) The total resistance is equal to the sum of the individual resistances.
    Explanation: In a series circuit, the total resistance is equal to the sum of the individual resistances, and the same current flows through each component.
48. What is the purpose of a raceway in electrical installations?
    a) To protect electrical devices from overload
    b) To provide a connection point for electrical wires
    c) To regulate the voltage in a circuit

d) To enclose and protect electrical conductors
Answer: d) To enclose and protect electrical conductors
Explanation: A raceway is used to enclose and protect electrical conductors, providing a pathway for the wires and ensuring safety and organization in electrical installations.

49. Which type of electrical panel is typically used as a secondary distribution panel, receiving power from the main panel?
a) Main panel
b) Subpanel
c) Load center
d) Distribution panel
Answer: b) Subpanel
Explanation: A subpanel is an additional electrical panel used as a secondary distribution panel, receiving power from the main panel and distributing it to specific areas or circuits.

50. What is the purpose of a current transformer (CT)?
a) To protect against overloads
b) To control the flow of current in a circuit
c) To regulate the voltage in a circuit
d) To measure and monitor electrical current
Answer: d) To measure and monitor electrical current
Explanation: A current transformer (CT) is used to measure and monitor electrical current in a circuit, providing a reduced current output proportional to the primary current.

51. Which of the following is an example of a type of electrical switch used to control motor loads?
a) Single-pole switch
b) Three-way switch
c) Toggle switch
d) Motor starter
Answer: d) Motor starter
Explanation: A motor starter is a type of electrical switch specifically designed to control motor loads, providing protection and control functions.

52. What is the purpose of a time-delay fuse?
a) To protect against overloads
b) To protect against short circuits
c) To regulate the voltage in a circuit
d) To provide a delay before interrupting the circuit
Answer: d) To provide a delay before interrupting the circuit
Explanation: Time-delay fuses are designed to provide a delay before interrupting the circuit, allowing temporary overloads to pass through without causing an immediate interruption.

53. Which of the following is a type of electrical wire that is suitable for direct burial in the ground?
a) THHN
b) NM
c) UF
d) MC
Answer: c) UF (Underground Feeder)
Explanation: UF (Underground Feeder) wire is specifically designed for direct burial in the ground, offering protection against moisture and other environmental factors.

54. What is the purpose of a ballast in a fluorescent lighting fixture?
a) To protect against overloads
b) To control the flow of current in the circuit
c) To regulate the voltage in the circuit
d) To provide the necessary voltage and current for the lamp to operate
Answer: d) To provide the necessary voltage and current for the lamp to operate
Explanation: A ballast is used in a fluorescent lighting fixture to provide the necessary voltage and current to start and operate the fluorescent lamp.

55. Which of the following is true about a delta-connected three-phase system?
    a) The line voltage is higher than the phase voltage.
    b) The line current is higher than the phase current.
    c) The phase voltage is higher than the line voltage.
    d) The phase current is higher than the line current.
    Answer: b) The line current is higher than the phase current.
    Explanation: In a delta-connected three-phase system, the line current is higher than the phase current due to the configuration of the connections.
56. What is the purpose of an AFCI (Arc-Fault Circuit Interrupter)?
    a) To protect against overloads
    b) To protect against short circuits
    c) To protect against ground faults
    d) To protect against arc faults
    Answer: d) To protect against arc faults
    Explanation: An AFCI is designed to detect and respond to arc faults, providing protection against potential fire hazards caused by arcing in electrical circuits.
57. Which of the following is a type of electrical switch that allows control of a single light fixture from multiple locations?
    a) Single-pole switch
    b) Three-way switch
    c) Four-way switch
    d) Dimmer switch
    Answer: c) Four-way switch
    Explanation: A four-way switch is used in conjunction with two three-way switches to allow control of a single light fixture from multiple locations.
58. What is the purpose of a service entrance conductor?
    a) To protect against overloads
    b) To control the flow of current in a circuit
    c) To regulate the voltage in a circuit
    d) To carry electrical power from the utility service to the main electrical panel
    Answer: d) To carry electrical power from the utility service to the main electrical panel
    Explanation: A service entrance conductor is responsible for carrying electrical power from the utility service to the main electrical panel of a building or structure.
59. Which of the following is a type of electrical wire that is commonly used for residential branch circuits?
    a) THHN
    b) UF
    c) NM
    d) AC
    Answer: c) NM (Non-Metallic)
    Explanation: NM (Non-Metallic) cable, also known as Romex, is commonly used for residential branch circuits, providing conductors and insulation in a single assembly.
60. What is the purpose of a bonding jumper in an electrical system?
    a) To control the flow of current in a circuit
    b) To regulate the voltage in a circuit
    c) To provide a connection point for electrical wires
    d) To establish electrical continuity between metal parts and ground
    Answer: d) To establish electrical continuity between metal parts and ground
    Explanation: A bonding jumper is used to establish electrical continuity between metal parts, such as electrical enclosures or equipment, and the grounding system, ensuring safety and effective grounding.
61. Which of the following is an example of a type of electrical hazard associated with static electricity?
    a) Ground fault
    b) Overcurrent
    c) Arc flash
    d) Electrostatic discharge

Answer: d) Electrostatic discharge
Explanation: Electrostatic discharge (ESD) refers to the sudden flow of electric charge between two objects, caused by static electricity buildup, which can be a potential electrical hazard.

62. What is the purpose of a ground fault sensor?
    a) To protect against overloads
    b) To protect against short circuits
    c) To sense and detect ground faults in electrical circuits
    d) To regulate the voltage in a circuit
    Answer: c) To sense and detect ground faults in electrical circuits
    Explanation: A ground fault sensor is designed to sense and detect ground faults in electrical circuits, providing a signal or initiating a protective action when a ground fault is detected.
63. Which of the following is a type of electrical wire that is commonly used for outdoor applications and direct burial?
    a) THHN
    b) NM
    c) UF
    d) MC
    Answer: c) UF (Underground Feeder)
    Explanation: UF (Underground Feeder) wire is specifically designed for outdoor applications and direct burial, providing protection against moisture and other environmental factors.
64. What is the purpose of a transformer's primary winding?
    a) To step up or step down voltage
    b) To regulate the current in a circuit
    c) To ground electrical systems
    d) To balance the load in a three-phase system
    Answer: a) To step up or step down voltage
    Explanation: The primary winding of a transformer is responsible for receiving the input voltage and is used to step up or step down the voltage, depending on the transformer's design.
65. Which of the following is a type of electrical switch that interrupts both the line and neutral conductors?
    a) Single-pole switch
    b) Double-pole switch
    c) Three-way switch
    d) Dimmer switch
    Answer: b) Double-pole switch
    Explanation: A double-pole switch is a type of electrical switch that interrupts both the line and neutral conductors, providing complete disconnection of the circuit.
66. What is the purpose of a bonding conductor in an electrical system?
    a) To protect against overloads
    b) To control the flow of current in a circuit
    c) To provide a connection point for electrical wires
    d) To establish electrical continuity between metal parts
    Answer: d) To establish electrical continuity between metal parts
    Explanation: A bonding conductor is used to establish electrical continuity between metal parts, such as electrical enclosures or equipment, ensuring safety and reducing the risk of electrical shock.
67. Which of the following is true about a parallel circuit?
    a) The total resistance increases as more components are added.
    b) The total current is the same throughout the circuit.
    c) The voltage across each component is the same.
    d) The total resistance is equal to the sum of the individual resistances.
    Answer: b) The total current is the same throughout the circuit.
    Explanation: In a parallel circuit, the total current is the sum of the currents flowing through each component, and it remains the same throughout the circuit.
68. What is the purpose of a ground electrode conductor?
    a) To protect against overloads
    b) To control the flow of current in a circuit

c) To provide a connection point for electrical wires
d) To establish a connection between the grounding electrode system and the grounding conductor
Answer: d) To establish a connection between the grounding electrode system and the grounding conductor
Explanation: A ground electrode conductor is used to establish a connection between the grounding electrode system, such as a ground rod, and the grounding conductor, ensuring effective grounding of electrical systems.

69. Which of the following is a type of electrical wire that is commonly used in conduit installations?
a) THHN
b) NM
c) UF
d) AC
Answer: a) THHN (Thermoplastic High Heat-resistant Nylon)
Explanation: THHN wire is commonly used in conduit installations, providing resistance to heat and suitable for both dry and wet locations.

70. What is the purpose of a transfer switch in a backup power system?
a) To protect against overloads
b) To control the flow of current in a circuit
c) To regulate the voltage in a circuit
d) To transfer the power source from the utility to the backup generator during a power outage
Answer: d) To transfer the power source from the utility to the backup generator during a power outage
Explanation: A transfer switch is used in a backup power system to transfer the power source from the utility to the backup generator automatically during a power outage, ensuring uninterrupted power supply.

71. Which of the following is an example of a type of electrical hazard associated with improper grounding?
a) Ground fault
b) Overcurrent
c) Arc flash
d) Electric shock
Answer: d) Electric shock
Explanation: Electric shock is an example of a type of electrical hazard that can occurI apologize for the incomplete response. Here's the continuation:

72. Which of the following is an example of a type of electrical hazard associated with improper grounding?
a) Ground fault
b) Overcurrent
c) Arc flash
d) Electric shock
Answer: d) Electric shock
Explanation: Electric shock is an example of a type of electrical hazard that can occur due to improper grounding, where an individual may come into contact with an energized component or conductor.

73. What is the purpose of a junction box in an electrical installation?
a) To protect against overloads
b) To control the flow of current in a circuit
c) To provide a connection point for electrical wires
d) To regulate the voltage in a circuit
Answer: c) To provide a connection point for electrical wires
Explanation: A junction box is used to provide a secure and accessible connection point for electrical wires, allowing for splicing, branching, or extending electrical circuits.

74. Which of the following is a type of electrical wire commonly used for high-voltage transmission lines?
a) THHN
b) NM
c) ACSR
d) MC
Answer: c) ACSR (Aluminum Conductor Steel Reinforced)
Explanation: ACSR wire is commonly used for high-voltage transmission lines due to its strength, low weight, and excellent conductivity.

75. What is the purpose of a transformer's secondary winding?
    a) To step up or step down voltage
    b) To regulate the current in a circuit
    c) To ground electrical systems
    d) To balance the load in a three-phase system
    Answer: a) To step up or step down voltage
    Explanation: The secondary winding of a transformer is responsible for delivering the stepped-up or stepped-down voltage as required by the load, based on the transformer's design.
76. Which of the following is true about a closed-loop control system?
    a) It does not require a feedback mechanism.
    b) It operates without any control devices.
    c) It uses feedback to continuously adjust and maintain a desired output.
    d) It does not respond to changes in the input.
    Answer: c) It uses feedback to continuously adjust and maintain a desired output.
    Explanation: A closed-loop control system utilizes feedback from the output to continuously adjust and regulate the system's input or control devices, ensuring the desired output is maintained.
77. What is the purpose of an isolating transformer?
    a) To protect against overloads
    b) To control the flow of current in a circuit
    c) To regulate the voltage in a circuit
    d) To electrically isolate one circuit from another
    Answer: d) To electrically isolate one circuit from another
    Explanation: An isolating transformer is designed to electrically isolate one circuit from another, preventing the transfer of electrical energy or potential differences between the circuits.
78. Which of the following is a type of electrical switch used to control the speed of motorized equipment?
    a) Single-pole switch
    b) Three-way switch
    c) Dimmer switch
    d) Motor controller switch
    Answer: c) Dimmer switch
    Explanation: A dimmer switch is used to control the intensity or speed of lighting fixtures or motorized equipment by adjusting the voltage or current supplied to the load.
79. What is the purpose of a load bank in electrical testing?
    a) To protect against overloads
    b) To control the flow of current in a circuit
    c) To regulate the voltage in a circuit
    d) To provide an artificial load for testing electrical systems or equipment
    Answer: d) To provide an artificial load for testing electrical systems or equipment
    Explanation: A load bank is a device used to provide an artificial load for testing electrical systems or equipment, allowing for the assessment of performance, capacity, and efficiency.
80. Which of the following is a type of electrical wire commonly used for underground installations?
    a) THHN
    b) NM
    c) UF
    d) AC
    Answer: c) UF (Underground Feeder)
    Explanation: UF wire is specifically designed for underground installations, offering protection against moisture, direct burial capability, and resistance to environmental factors.
81. What is the purpose of a shunt trip in a circuit breaker?
    a) To protect against overloads
    b) To control the flow of current in a circuit
    c) To regulate the voltage in a circuit
    d) To remotely trip the circuit breaker under certain conditions
    Answer: d) To remotely trip the circuit breaker under certain conditions

Explanation: A shunt trip is an accessory used in circuit breakers to provide remote tripping capability, allowing the circuit breaker to be tripped remotely under specific conditions or in emergency situations.

82. Which of the following is an example of a type of electrical switch used to control lighting fixtures from multiple locations?
    a) Single-pole switch
    b) Three-way switch
    c) Four-way switch
    d) Toggle switch
    Answer: c) Four-way switch
    Explanation: A four-way switch is used in conjunction with two three-way switches to control lighting fixtures from multiple locations.
83. What is the purpose of a ground fault circuit interrupter (GFCI)?
    a) To protect against overloads
    b) To protect against short circuits
    c) To protect against ground faults
    d) To regulate the voltage in a circuit
    Answer: c) To protect against ground faults
    Explanation: A ground fault circuit interrupter (GFCI) is designed to quickly interrupt the circuit when it detects a ground fault, providing protection against electrical shock.
84. Which of the following is a type of electrical wire that is commonly used for residential branch circuits?
    a) THHN
    b) NM
    c) UF
    d) AC
    Answer: b) NM (Non-Metallic)
    Explanation: NM (Non-Metallic) cable, also known as Romex, is commonly used for residential branch circuits, providing conductors and insulation in a single assembly.
85. What is the purpose of a transformer's tap changer?
    a) To protect against overloads
    b) To control the flow of current in a circuit
    c) To regulate the voltage in a circuit
    d) To adjust the transformer's output voltage based on load conditions
    Answer: d) To adjust the transformer's output voltage based on load conditions
    Explanation: A tap changer is used to adjust the output voltage of a transformer based on load conditions, ensuring the desired voltage level is maintained.
86. Which of the following is an example of a type of electrical hazard associated with improper wiring connections?
    a) Ground fault
    b) Overcurrent
    c) Arc flash
    d) Short circuit
    Answer: c) Arc flash
    Explanation: Arc flash is an example of a type of electrical hazard that can occur due to improper wiring connections, resulting in an explosive release of energy that can cause severe burns and injuries.
87. What is the purpose of a capacitor in an AC circuit?
    a) To store and release electrical energy
    b) To control the flow of current
    c) To regulate the voltage
    d) To ground the circuit
    Answer: a) To store and release electrical energy
    Explanation: Capacitors are used in AC circuits to store electrical energy and release it when needed, such as during changes in voltage or in timing circuits.
88. Which of the following is a type of electrical switch commonly used in motor control circuits?
    a) Toggle switch
    b) Selector switch

c) Push-button switch
d) Dimmer switch
Answer: b) Selector switch
Explanation: Selector switches are commonly used in motor control circuits to switch between different control modes or operations, such as forward or reverse rotation.

89. What is the purpose of a ground fault locator tool?
a) To protect against overloads
b) To measure voltage in a circuit
c) To locate the source of a ground fault
d) To test the continuity of a circuit
Answer: c) To locate the source of a ground fault
Explanation: A ground fault locator tool is used to identify and locate the source of a ground fault in an electrical circuit, helping electricians troubleshoot and fix the problem.

90. Which of the following is an example of a type of electrical wire that is suitable for wet locations and direct burial?
a) THHN
b) NM
c) UF
d) MC
Answer: c) UF (Underground Feeder)
Explanation: UF wire is designed for wet locations and direct burial, making it suitable for underground installations where moisture resistance is required.

91. What is the purpose of a phase sequence indicator?
a) To protect against overloads
b) To control the flow of current in a circuit
c) To regulate the voltage in a circuit
d) To identify the order of phases in a three-phase system
Answer: d) To identify the order of phases in a three-phase system
Explanation: A phase sequence indicator is used to determine the correct order of phases in a three-phase electrical system, ensuring proper installation and operation of equipment.

92. What is the purpose of a time-delay relay in an electrical circuit?
a) To protect against overloads
b) To control the flow of current in a circuit
c) To regulate the voltage in a circuit
d) To introduce a time delay before activating or deactivating a circuit
Answer: d) To introduce a time delay before activating or deactivating a circuit
Explanation: A time-delay relay is used to introduce a specific time delay before activating or deactivating a circuit, allowing for control and coordination of electrical operations.

93. Which of the following is a type of electrical wire commonly used for grounding applications?
a) THHN
b) NM
c) GFCI
d) Bare copper
Answer: d) Bare copper
Explanation: Bare copper wire is commonly used for grounding applications, providing a reliable and low-resistance path for fault currents to flow safely into the ground.

94. What is the purpose of a photovoltaic (PV) system in electrical installations?
a) To protect against overloads
b) To control the flow of current in a circuit
c) To generate electricity from sunlight
d) To regulate the voltage in a circuit
Answer: c) To generate electricity from sunlight
Explanation: A photovoltaic (PV) system is used to generate electricity from sunlight, converting solar energy into usable electrical power.

95. Which of the following is an example of a type of electrical switch used to control the speed of an electric motor?
    a) Single-pole switch
    b) Three-way switch
    c) Rheostat
    d) Pressure switch
    Answer: c) Rheostat
    Explanation: A rheostat is an example of an electrical switch used to control the speed or intensity of an electric motor by varying the resistance in the circuit.
96. What is the purpose of a ground fault locator tool?
    a) To protect against overloads
    b) To measure voltage in a circuit
    c) To locate the source of a ground fault
    d) To test the continuity of a circuit
    Answer: c) To locate the source of a ground fault
    Explanation: A ground fault locator tool is used to identify and locate the source of a ground fault in an electrical circuit, helping electricians troubleshoot and fix the problem.
97. Which of the following is an example of a type of electrical enclosure used for protecting electrical equipment against environmental factors?
    a) Junction box
    b) Outlet box
    c) Weatherproof box
    d) Pull box
    Answer: c) Weatherproof box
    Explanation: A weatherproof box is specifically designed to protect electrical equipment from environmental factors such as rain, snow, and dust.
98. What is the purpose of a ground fault protection device?
    a) To protect against overloads
    b) To protect against short circuits
    c) To protect against ground faults
    d) To regulate the voltage in a circuit
    Answer: c) To protect against ground faults
    Explanation: A ground fault protection device, such as a ground fault circuit interrupter (GFCI), is designed to quickly interrupt the circuit when it detects a ground fault, providing protection against electrical shock.
99. Which of the following is a type of electrical wire that is suitable for use in conduit installations?
    a) THHN
    b) NM
    c) UF
    d) AC
    Answer: a) THHN (Thermoplastic High Heat-resistant Nylon)
    Explanation: THHN wire is commonly used in conduit installations due to its resistance to heat and its suitability for dry and wet locations.
100. What is the purpose of a phase sequence meter in a three-phase electrical system?
    a) To protect against overloads
    b) To control the flow of current in a circuit
    c) To regulate the voltage in a circuit
    d) To determine the correct sequence of phases
    Answer: d) To determine the correct sequence of phases
    Explanation: A phase sequence meter is used to determine the correct sequence of phases in a three-phase electrical system, ensuring proper installation and operation of equipment.
101. Which of the following is an example of a type of electrical hazard associated with improper insulation?
    a) Ground fault
    b) Overcurrent
    c) Arc flash
    d) Electric shock

Answer: d) Electric shock
Explanation: Electric shock can occur as a result of improper insulation, where a person comes into contact with an exposed or inadequately insulated conductor.

102. What is the purpose of a motor overload protection device?
a) To protect against overloads
b) To control the flow of current in a circuit
c) To regulate the voltage in a circuit
d) To provide a connection point for electrical wires
Answer: a) To protect against overloads
Explanation: A motor overload protection device is designed to protect a motor against excessive current and overloads that can lead to motor damage or failure.

103. Which of the following is a type of electrical wire that is commonly used for high-temperature applications?
a) THHN
b) NM
c) XHHW
d) MC
Answer: c) XHHW (Cross-Linked Polyethylene High Heat-Resistant)
Explanation: XHHW wire is designed for high-temperature applications and has excellent resistance to heat, making it suitable for various environments.

104. What is the purpose of a transformer's core?
a) To step up or step down voltage
b) To regulate the current in a circuit
c) To ground electrical systems
d) To provide a path for magnetic flux and enhance transformer efficiency
Answer: d) To provide a path for magnetic flux and enhance transformer efficiency
Explanation: The transformer's core is made of high-permeability material and provides a closed path for the magnetic flux, which helps enhance the efficiency of the transformer.

105. Which of the following is an example of a type of electrical switch used for emergency shutdown of equipment?
a) Single-pole switch
b) Emergency stop switch
c) Selector switch
d) Pressure switch
Answer: b) Emergency stop switch
Explanation: An emergency stop switch, also known as an emergency stop button or E-stop, is specifically designed for immediate shutdown of equipment in emergency situations.

106. What is the purpose of a ground fault alarm system?
a) To protect against overloads
b) To control the flow of current in a circuit
c) To alert when a ground fault occurs in an electrical system
d) To regulate the voltage in a circuit
Answer: c) To alert when a ground fault occurs in an electrical system
Explanation: A ground fault alarm system is designed to detect and signal the presence of a ground fault in an electrical system, providing an alert to prompt necessary actions.

107. What is the purpose of a motor starter in an electrical system?
a) To protect against overloads and short circuits
b) To control the flow of current in a circuit
c) To regulate the voltage in a circuit
d) To start and stop an electric motor
Answer: d) To start and stop an electric motor
Explanation: A motor starter is an electrical device used to start and stop an electric motor, providing control and protection functions.

108. Which of the following is a type of electrical wire that is typically used for wiring residential outlets and lighting circuits?
a) THHN

b) NM
c) UF
d) AC
Answer: b) NM (Non-Metallic)
Explanation: NM (Non-Metallic) cable, also known as Romex, is commonly used for wiring residential outlets and lighting circuits, providing both conductors and insulation in a single assembly.

109.What is the purpose of a ground fault locator tool?
a) To protect against overloads
b) To measure voltage in a circuit
c) To locate the source of a ground fault
d) To test the continuity of a circuit
Answer: c) To locate the source of a ground fault
Explanation: A ground fault locator tool is used to identify and locate the source of a ground fault in an electrical circuit, assisting in troubleshooting and repair.

110.Which of the following is an example of a type of electrical switch used for motor control applications, allowing for forward and reverse motor operation?
a) Single-pole switch
b) Three-way switch
c) Reversing switch
d) Selector switch
Answer: c) Reversing switch
Explanation: A reversing switch is used for motor control applications, allowing for forward and reverse motor operation by changing the direction of current flow through the motor windings.

111.What is the purpose of a disconnect switch in an electrical system?
a) To protect against overloads
b) To control the flow of current in a circuit
c) To provide a connection point for electrical wires
d) To manually disconnect power from a circuit or equipment
Answer: d) To manually disconnect power from a circuit or equipment
Explanation: A disconnect switch is used to manually interrupt or disconnect power from a circuit or equipment, providing a means for safe maintenance and repair activities.

112.What is the purpose of a step-up transformer?
a) To protect against overloads
b) To control the flow of current in a circuit
c) To regulate the voltage in a circuit
d) To increase the voltage level
Answer: d) To increase the voltage level
Explanation: A step-up transformer is used to increase the voltage level, typically used in power transmission to step up the voltage for efficient long-distance transmission.

113.Which of the following is a type of electrical wire that is commonly used for underground feeder circuits?
a) THHN
b) NM
c) UF
d) AC
Answer: c) UF (Underground Feeder)
Explanation: UF (Underground Feeder) wire is specifically designed for underground feeder circuits, providing protection against moisture and other environmental factors.

114.What is the purpose of a control transformer in an electrical system?
a) To protect against overloads
b) To control the flow of current in a circuit
c) To regulate the voltage in a circuit
d) To provide low-voltage control power for control circuits
Answer: d) To provide low-voltage control power for control circuits

Explanation: A control transformer is used to provide low-voltage control power for control circuits in an electrical system, ensuring safe and reliable operation of control devices.

115. Which of the following is an example of a type of electrical switch used for motor control applications, allowing for speed adjustment?
a) Single-pole switch
b) Three-way switch
c) Dimmer switch
d) Selector switch
Answer: c) Dimmer switch
Explanation: While dimmer switches are commonly used for lighting control, they can also be employed for speed adjustment of certain types of motors.

116. What is the purpose of a voltage regulator in an electrical system?
a) To protect against overloads
b) To control the flow of current in a circuit
c) To regulate the voltage at a constant level
d) To provide a connection point for electrical wires
Answer: c) To regulate the voltage at a constant level
Explanation: A voltage regulator is used to regulate the voltage at a constant level, ensuring a stable power supply within a specific voltage range.

117. What is the purpose of a busway in an electrical system?
a) To protect against overloads
b) To control the flow of current in a circuit
c) To distribute electrical power within a building or facility
d) To regulate the voltage in a circuit
Answer: c) To distribute electrical power within a building or facility
Explanation: A busway is used to distribute electrical power within a building or facility, providing a safe and efficient means of transmitting electricity to various loads.

118. Which of the following is a type of electrical wire that is commonly used for outdoor and direct burial applications?
a) THHN
b) NM
c) UF
d) AC
Answer: c) UF (Underground Feeder)
Explanation: UF (Underground Feeder) wire is specifically designed for outdoor and direct burial applications, offering protection against moisture and environmental factors.

119. What is the purpose of a power factor correction capacitor in an electrical system?
a) To protect against overloads
b) To control the flow of current in a circuit
c) To improve the power factor and increase energy efficiency
d) To regulate the voltage in a circuit
Answer: c) To improve the power factor and increase energy efficiency
Explanation: A power factor correction capacitor is used to improve the power factor of an electrical system, reducing reactive power and increasing energy efficiency.

120. Which of the following is an example of a type of electrical switch used for controlling lighting fixtures from multiple locations?
a) Single-pole switch
b) Three-way switch
c) Four-way switch
d) Toggle switch
Answer: c) Four-way switch
Explanation: A four-way switch is used in combination with two three-way switches to control lighting fixtures from multiple locations.

121. What is the purpose of a power quality analyzer in an electrical system?
a) To protect against overloads

b) To control the flow of current in a circuit
c) To measure and analyze electrical parameters for evaluating power quality
d) To regulate the voltage in a circuit
Answer: c) To measure and analyze electrical parameters for evaluating power quality
Explanation: A power quality analyzer is used to measure and analyze electrical parameters such as voltage, current, harmonics, and other factors to evaluate and assess the power quality in an electrical system.

122. What is the purpose of a surge protective device (SPD)?
a) To protect against overloads
b) To control the flow of current in a circuit
c) To regulate the voltage in a circuit
d) To protect against voltage surges and transients
Answer: d) To protect against voltage surges and transients
Explanation: A surge protective device (SPD) is designed to divert excess voltage and protect against voltage surges and transients caused by lightning strikes, switching operations, or other events.

123. Which of the following is a type of electrical wire that is commonly used for high-temperature applications and in hazardous locations?
a) THHN
b) NM
c) SE
d) MC-HL
Answer: d) MC-HL (Metal-Clad Hazardous Location)
Explanation: MC-HL wire is specifically designed for high-temperature applications and use in hazardous locations, offering enhanced protection against fire and explosion risks.

124. What is the purpose of a ground-fault protection relay?
a) To protect against overloads
b) To control the flow of current in a circuit
c) To regulate the voltage in a circuit
d) To sense and detect ground faults and initiate protective actions
Answer: d) To sense and detect ground faults and initiate protective actions
Explanation: A ground-fault protection relay is a device that senses and detects ground faults in electrical circuits and initiates protective actions, such as tripping a circuit breaker, to prevent electrical shock hazards.

125. Which of the following is an example of a type of electrical switch used to control the operation of a motor based on temperature conditions?
a) Single-pole switch
b) Three-way switch
c) Pressure switch
d) Selector switch
Answer: c) Pressure switch
Explanation: A pressure switch is commonly used to control the operation of a motor based on temperature conditions, such as turning on a cooling fan when the temperature exceeds a certain threshold.

126. What is the purpose of a lockout/tagout (LOTO) device in electrical maintenance?
a) To protect against overloads
b) To control the flow of current in a circuit
c) To regulate the voltage in a circuit
d) To isolate and de-energize electrical equipment during maintenance or servicing
Answer: d) To isolate and de-energize electrical equipment during maintenance or servicing
Explanation: A lockout/tagout (LOTO) device is used to isolate and de-energize electrical equipment during maintenance or servicing, preventing unintentional energization and ensuring the safety of personnel working on the equipment.

127. What is the purpose of a motor contactor in an electrical system?
a) To protect against overloads
b) To control the flow of current in a circuit
c) To regulate the voltage in a circuit
d) To switch power to a motor and control its operation

Answer: d) To switch power to a motor and control its operation
Explanation: A motor contactor is an electrical device used to switch power to a motor and control its operation, allowing for remote control and protection of the motor.

128. Which of the following is a type of electrical wire that is commonly used for high-voltage distribution lines?
a) THHN
b) NM
c) ACSR
d) MC
Answer: c) ACSR (Aluminum Conductor Steel Reinforced)
Explanation: ACSR wire is commonly used for high-voltage distribution lines due to its strength, low weight, and high conductivity.

129. What is the purpose of a photovoltaic inverter in a solar power system?
a) To protect against overloads
b) To control the flow of current in a circuit
c) To regulate the voltage in a circuit
d) To convert the DC power generated by solar panels into AC power
Answer: d) To convert the DC power generated by solar panels into AC power
Explanation: A photovoltaic inverter is used in a solar power system to convert the DC power generated by solar panels into AC power that can be used by electrical loads or fed back into the grid.

130. Which of the following is an example of a type of electrical switch used for controlling the speed of a ceiling fan?
a) Single-pole switch
b) Three-way switch
c) Dimmer switch
d) Pull chain switch
Answer: d) Pull chain switch
Explanation: A pull chain switch is commonly used for controlling the speed of a ceiling fan, allowing for multiple speed settings by pulling a chain attached to the switch.

131. What is the purpose of an electrical bonding jumper in a grounding system?
a) To protect against overloads
b) To control the flow of current in a circuit
c) To provide a connection point for electrical wires
d) To ensure electrical continuity between metal parts for effective grounding
Answer: d) To ensure electrical continuity between metal parts for effective grounding
Explanation: An electrical bonding jumper is used to ensure electrical continuity between metal parts, such as electrical enclosures or equipment, for effective grounding and to minimize voltage potential differences.

132. What is the purpose of a digital multimeter (DMM) in electrical troubleshooting?
a) To protect against overloads
b) To control the flow of current in a circuit
c) To measure voltage, current, and resistance in electrical circuits
d) To regulate the voltage in a circuit
Answer: c) To measure voltage, current, and resistance in electrical circuits
Explanation: A digital multimeter (DMM) is a versatile tool used for electrical troubleshooting and measurement, allowing electricians to measure voltage, current, and resistance in electrical circuits.

133. Which of the following is a type of electrical wire that is commonly used for indoor residential wiring?
a) THHN
b) NM
c) UF
d) AC
Answer: b) NM (Non-Metallic)
Explanation: NM (Non-Metallic) cable, also known as Romex, is commonly used for indoor residential wiring, providing both conductors and insulation in a single assembly.

134. What is the purpose of an electrical ground rod?
a) To protect against overloads
b) To control the flow of current in a circuit

c) To provide a connection point for electrical wires
d) To establish an effective grounding system for electrical equipment and systems
Answer: d) To establish an effective grounding system for electrical equipment and systems
Explanation: An electrical ground rod is driven into the earth to establish an effective grounding system, providing a safe path for electrical fault currents and preventing electrical hazards.

135. Which of the following is an example of a type of electrical switch used for controlling lighting fixtures from multiple locations?
a) Single-pole switch
b) Three-way switch
c) Four-way switch
d) Toggle switch
Answer: c) Four-way switch
Explanation: A four-way switch is used in conjunction with two three-way switches to control lighting fixtures from multiple locations.

136. What is the purpose of a time clock in an electrical system?
a) To protect against overloads
b) To control the flow of current in a circuit
c) To regulate the voltage in a circuit
d) To automatically control the timing of electrical operations or devices
Answer: d) To automatically control the timing of electrical operations or devices
Explanation: A time clock is used to automatically control the timing of electrical operations or devices, allowing for scheduled on/off switching or timed functions.

137. What is the purpose of a ground fault circuit interrupter (GFCI) receptacle tester?
a) To protect against overloads
b) To control the flow of current in a circuit
c) To test the functionality of GFCI receptacles and identify wiring faults
d) To regulate the voltage in a circuit
Answer: c) To test the functionality of GFCI receptacles and identify wiring faults
Explanation: A GFCI receptacle tester is used to test the functionality of GFCI receptacles and identify wiring faults, ensuring they are operating correctly and providing proper protection.

138. Which of the following is a type of electrical wire that is commonly used for overhead transmission lines?
a) THHN
b) NM
c) ACSR
d) MC
Answer: c) ACSR (Aluminum Conductor Steel Reinforced)
Explanation: ACSR wire is commonly used for overhead transmission lines due to its high strength, low weight, and good conductivity.

139. What is the purpose of a pulse oximeter in an electrical system?
a) To protect against overloads
b) To control the flow of current in a circuit
c) To measure and monitor blood oxygen levels
d) To regulate the voltage in a circuit
Answer: c) To measure and monitor blood oxygen levels
Explanation: A pulse oximeter is a medical device that measures and monitors blood oxygen levels, helping to assess a patient's respiratory status.

140. Which of the following is an example of a type of electrical switch used for controlling the speed of a motor?
a) Single-pole switch
b) Three-way switch
c) Variable frequency drive (VFD)
d) Selector switch
Answer: c) Variable frequency drive (VFD)
Explanation: A variable frequency drive (VFD) is an electrical device used to control the speed of a motor by varying the frequency and voltage supplied to it.

141. What is the purpose of a ground fault locator tool?
a) To protect against overloads
b) To measure voltage in a circuit
c) To locate the source of a ground fault
d) To test the continuity of a circuit
Answer: c) To locate the source of a ground fault
Explanation: A ground fault locator tool is used to identify and locate the source of a ground fault in an electrical circuit, aiding in troubleshooting and repair.

142. What is the purpose of a power factor correction device in an electrical system?
a) To protect against overloads
b) To control the flow of current in a circuit
c) To regulate the voltage in a circuit
d) To improve the power factor and increase energy efficiency
Answer: d) To improve the power factor and increase energy efficiency
Explanation: A power factor correction device is used to improve the power factor of an electrical system, reducing reactive power and increasing energy efficiency.

143. Which of the following is a type of electrical wire commonly used for overhead service entrance wiring?
a) THHN
b) NM
c) SE
d) MC
Answer: c) SE (Service Entrance)
Explanation: SE (Service Entrance) cable is commonly used for overhead service entrance wiring, providing insulation and protection for conductors.

144. What is the purpose of a phase rotation meter in an electrical system?
a) To protect against overloads
b) To control the flow of current in a circuit
c) To measure and determine the correct rotation sequence of three-phase power
d) To regulate the voltage in a circuit
Answer: c) To measure and determine the correct rotation sequence of three-phase power
Explanation: A phase rotation meter is used to measure and determine the correct rotation sequence of three-phase power, ensuring proper installation and operation of electrical equipment.

145. Which of the following is an example of a type of electrical switch used for controlling lighting fixtures with adjustable brightness?
a) Single-pole switch
b) Three-way switch
c) Dimmer switch
d) Selector switch
Answer: c) Dimmer switch
Explanation: A dimmer switch is used for controlling lighting fixtures with adjustable brightness, allowing users to vary the intensity of the light.

146. What is the purpose of a lockout/tagout (LOTO) procedure in electrical work?
a) To protect against overloads
b) To control the flow of current in a circuit
c) To regulate the voltage in a circuit
d) To ensure the safe de-energization and isolation of electrical equipment during maintenance or repair
Answer: d) To ensure the safe de-energization and isolation of electrical equipment during maintenance or repair
Explanation: A lockout/tagout (LOTO) procedure is used to ensure the safe de-energization and isolation of electrical equipment during maintenance or repair, preventing unintended energization and protecting workers from electrical hazards.

147. What is the purpose of a ground fault circuit interrupter (GFCI) in an electrical system?
a) To protect against overloads
b) To control the flow of current in a circuit
c) To regulate the voltage in a circuit

d) To provide protection against electrical shock
Answer: d) To provide protection against electrical shock
Explanation: A ground fault circuit interrupter (GFCI) is designed to monitor the flow of current and quickly interrupt the circuit if it detects a ground fault, providing protection against electrical shock.

148. Which of the following is a type of electrical wire that is commonly used for marine applications?
a) THHN
b) NM
c) SJOOW
d) AC
Answer: c) SJOOW
Explanation: SJOOW wire is commonly used for marine applications due to its resistance to water, oil, and other environmental factors.

149. What is the purpose of a current transformer (CT) in an electrical system?
a) To protect against overloads
b) To control the flow of current in a circuit
c) To regulate the voltage in a circuit
d) To measure and monitor electrical current
Answer: d) To measure and monitor electrical current
Explanation: A current transformer (CT) is used to measure and monitor electrical current in a circuit, providing a reduced current output proportional to the primary current.

150. Which of the following is an example of a type of electrical switch used for controlling the operation of a motor based on liquid or fluid levels?
a) Single-pole switch
b) Three-way switch
c) Float switch
d) Selector switch
Answer: c) Float switch
Explanation: A float switch is commonly used for controlling the operation of a motor based on liquid or fluid levels, activating or deactivating the motor as the liquid reaches a certain level.

151. What is the purpose of an electrical conduit in an electrical installation?
a) To protect against overloads
b) To control the flow of current in a circuit
c) To regulate the voltage in a circuit
d) To provide a protective pathway for electrical wires
Answer: d) To provide a protective pathway for electrical wires
Explanation: An electrical conduit is used to provide a protective pathway for electrical wires, offering physical protection and helping to prevent damage, exposure, and interference.

152. What is the purpose of a ground fault circuit interrupter (GFCI) in an electrical system?
a) To protect against short circuits
b) To protect against voltage fluctuations
c) To provide grounding for electrical equipment
d) To provide protection against electrical shock hazards

Answer: d) To provide protection against electrical shock hazards.
Explanation: A ground fault circuit interrupter (GFCI) continuously monitors the current flowing through a circuit. If it detects an imbalance between the hot and neutral conductors, indicating a ground fault, it quickly interrupts the circuit to prevent electrical shocks.

153. What is the purpose of a transformer in an electrical system?
a) To convert AC to DC power
b) To regulate voltage levels
c) To protect against overcurrents
d) To switch between power sources

Answer: b) To regulate voltage levels.
Explanation: A transformer is used to step up or step down voltage levels in an electrical system. It consists of primary and secondary windings that are magnetically coupled, allowing for efficient voltage transformation.

154. What is the purpose of a circuit breaker in an electrical system?
a) To generate electrical power
b) To control motor speed
c) To protect against overloads and short circuits
d) To regulate voltage levels

Answer: c) To protect against overloads and short circuits.
Explanation: A circuit breaker is a protective device that automatically interrupts the flow of current when it exceeds a predetermined level or in the event of a short circuit. It helps protect the electrical system and connected devices from damage.

155. What is the purpose of an electrical bonding jumper in a grounding system?
a) To measure electrical conductivity
b) To establish electrical continuity and bonding
c) To reduce electrical resistance
d) To generate grounding voltage

Answer: b) To establish electrical continuity and bonding.
Explanation: An electrical bonding jumper is used to connect metal parts, such as electrical enclosures or equipment, to establish electrical continuity and bonding. This ensures effective grounding, reduces the risk of electrical shock hazards, and equalizes potentials between metal components.

156. What is the purpose of an ammeter in an electrical system?
a) To measure voltage
b) To measure resistance
c) To measure electric current
d) To measure power factor

Answer: c) To measure electric current.
Explanation: An ammeter, also known as an ampere meter, is used to measure the electric current flowing through a circuit. It is connected in series with the circuit and provides a direct reading of the current in amperes.

157. What is the purpose of a three-way switch in a lighting circuit?
a) To control the brightness of the light
b) To control the color temperature of the light
c) To control the light from three different locations
d) To control the light from two different locations

Answer: d) To control the light from two different locations.
Explanation: A three-way switch is used in a lighting circuit to control the light from two different locations. It allows the user to turn the light on or off from either switch position, providing convenience and flexibility in lighting control.

158. What is the purpose of an arc-fault circuit interrupter (AFCI) in an electrical system?
a) To protect against voltage surges
b) To protect against short circuits
c) To protect against ground faults
d) To protect against electrical fires caused by arc faults

Answer: d) To protect against electrical fires caused by arc faults.
Explanation: An arc-fault circuit interrupter (AFCI) is a protective device that detects and interrupts abnormal arcing conditions in an electrical system. It helps prevent potential fire hazards by quickly interrupting the circuit upon detecting arcing faults, such as those caused by damaged or deteriorating wiring.

159.What is the purpose of a junction box in an electrical installation?
- a) To connect electrical devices together
- b) To provide grounding for electrical circuits
- c) To protect electrical connections and contain sparks or heat
- d) To switch between different power sources

Answer: c) To protect electrical connections and contain sparks or heat.
Explanation: A junction box is used in an electrical installation to provide a safe and accessible enclosure for electrical connections. It helps protect the connections from damage, prevents accidental contact, and contains any sparks or heat generated within the box.

160.Which of the following is an example of a conductor material commonly used in electrical wiring?
- a) Copper
- b) Plastic
- c) Wood
- d) Glass

Answer: a) Copper
Detailed Explanation: Copper is a commonly used conductor material in electrical wiring due to its excellent electrical conductivity. It is preferred for its low resistance, durability, and ability to carry high current loads effectively.

161.What is the primary purpose of a ground fault circuit interrupter (GFCI)?
- a) To protect against overvoltage
- b) To prevent electrical shock
- c) To regulate current flow
- d) To detect short circuits

Answer: b) To prevent electrical shock
Detailed Explanation: A ground fault circuit interrupter (GFCI) is designed to detect even small imbalances in electrical current and quickly shut off power to prevent electric shock. It provides protection against ground faults and is typically used in areas where water is present, such as bathrooms and kitchens.

162.What is the minimum wire size allowed for a 20-ampere branch circuit?
- a) 14 AWG
- b) 12 AWG
- c) 10 AWG
- d) 8 AWG

Answer: b) 12 AWG
Detailed Explanation: According to the National Electrical Code (NEC), a 20-ampere branch circuit should be wired with a minimum of 12 AWG (American Wire Gauge) copper wire. This wire size is capable of handling the current without excessive voltage drop or overheating.

163.Which of the following is the standard voltage for residential electrical systems in the United States?
- a) 120 volts
- b) 240 volts
- c) 480 volts
- d) 600 volts

Answer: a) 120 volts
Detailed Explanation: The standard voltage for residential electrical systems in the United States is 120 volts. This voltage is used for general lighting, small appliances, and most household outlets. Higher voltages, such as 240 volts, are typically used for larger appliances and specialized equipment.

164.When installing electrical wiring in a damp location, which type of wiring method would be most appropriate?
- a) Non-metallic sheathed cable (NM)
- b) Armored cable (AC)
- c) Rigid metal conduit (RMC)
- d) Liquidtight flexible metal conduit (LFMC)

Answer: d) Liquidtight flexible metal conduit (LFMC)
Detailed Explanation: In damp locations, it is important to use wiring methods that provide protection against moisture. Liquidtight flexible metal conduit (LFMC) is a suitable choice as it is designed to be resistant to liquids, including water. It

provides a high level of protection against moisture and is commonly used in areas such as basements, garages, and outdoor installations.

165. Which of the following is an example of a single-pole switch?

a) Three-way switch
b) Four-way switch
c) Dimmer switch
d) Toggle switch

Answer: d) Toggle switch

Detailed Explanation: A single-pole switch is the most basic type of switch used to control a light or electrical device from a single location. A toggle switch is a common example of a single-pole switch, where a flip of the toggle either turns the device on or off.

166. What is the purpose of a junction box in an electrical system?

a) To support light fixtures
b) To connect electrical wires
c) To protect against electrical surges
d) To provide grounding for circuits

Answer: b) To connect electrical wires

Detailed Explanation: A junction box is used to securely connect and protect electrical wires in an electrical system. It provides a safe and enclosed space where wires can be joined together using wire nuts or other approved connectors. Junction boxes also help prevent accidental contact with live wires and contribute to the overall safety of the electrical installation.

167. Which of the following is the correct formula to calculate electrical power?

a) Power = Voltage x Current
b) Power = Current / Resistance
c) Power = Voltage / Current
d) Power = Resistance / Current

Answer: a) Power = Voltage x Current

Detailed Explanation: The formula to calculate electrical power is Power = Voltage x Current. Power is measured in watts (W) and is the product of the voltage (V) applied across a device or circuit and the current (I) flowing through it. This formula demonstrates the relationship between voltage, current, and power in an electrical system.

168. What is the purpose of a circuit breaker in an electrical system?

a) To regulate voltage levels
b) To measure electrical resistance
c) To control the flow of current
d) To provide electrical grounding

Answer: c) To control the flow of current

Detailed Explanation: A circuit breaker is a protective device used to control and interrupt the flow of electrical current in a circuit. It is designed169. What is the purpose of a junction box in an electrical system?

a) To support light fixtures
b) To connect electrical wires
c) To protect against electrical surges
d) To provide grounding for circuits

Answer: b) To connect electrical wires

Detailed Explanation: A junction box is used to securely connect and protect electrical wires in an electrical system. It provides a safe and enclosed space where wires can be joined together using wire nuts or other approved connectors. Junction boxes also help prevent accidental contact with live wires and contribute to the overall safety of the electrical installation.

170. Which of the following is the correct formula to calculate electrical power?

a) Power = Voltage x Current
b) Power = Current / Resistance
c) Power = Voltage / Current
d) Power = Resistance / Current

Answer: a) Power = Voltage x Current

Detailed Explanation: The formula to calculate electrical power is Power = Voltage x Current. Power is measured in watts

(W) and is the product of the voltage (V) applied across a device or circuit and the current (I) flowing through it. This formula demonstrates the relationship between voltage, current, and power in an electrical system.

171. What is the purpose of a circuit breaker in an electrical system?
   a) To regulate voltage levels
   b) To measure electrical resistance
   c) To control the flow of current
   d) To provide electrical grounding

Answer: c) To control the flow of current
Detailed Explanation: A circuit breaker is a protective device used to control and interrupt the flow of electrical current in a circuit. It is designed to automatically open the circuit and stop the flow of current when an overload or short circuit occurs. By doing so, it helps protect the electrical system and prevent damage to equipment and potential hazards such as fires.

172. Which of the following is a unit of electrical resistance?
   a) Ampere (A)
   b) Volt (V)
   c) Ohm (Ω)
   d) Watt (W)

Answer: c) Ohm (Ω)
Detailed Explanation: The unit of electrical resistance is the ohm (Ω). It is named after the German physicist Georg Simon Ohm and is represented by the Greek letter omega (Ω). Resistance is a measure of how much an electrical component or material opposes the flow of electric current.

173. What does the term "grounding" refer to in electrical systems?
   a) Connecting electrical devices to the Earth
   b) Establishing a reference voltage level
   c) Preventing electrical shock hazards
   d) Protecting against power surges

Answer: a) Connecting electrical devices to the Earth
Detailed Explanation: Grounding in electrical systems refers to the process of connecting electrical devices, equipment, and structures to the Earth or a grounding electrode system. This connection serves several purposes, including providing a safe path for electrical faults, reducing electrical noise, and ensuring the effectiveness of overcurrent protection devices.

174. Which of the following symbols represents a resistor in an electrical circuit diagram?
   a) A circle
   b) A square
   c) A triangle
   d) A zigzag line

Answer: d) A zigzag line
Detailed Explanation: In electrical circuit diagrams, a zigzag line is commonly used to represent a resistor. A resistor is an electronic component that restricts the flow of electric current, dissipating electrical energy in the form of heat. It is often used to control the amount of current or voltage in a circuit.

175. What is the purpose of a transformer in an electrical system?
   a) To convert AC voltage to DC voltage
   b) To regulate electrical frequency
   c) To step up or step down voltage
   d) To provide electrical isolation

Answer: c) To step up or step down voltage
Detailed Explanation: A transformer is an electrical device used to transfer electrical energy between two or more circuits through electromagnetic induction. Its primary function is to step up (increase) or step down (decrease) the voltage levels between the input and output circuits. Transformers are commonly used in power distribution systems to adjust voltage levels for efficient transmission and utilization.

176. What is the purpose of a three-way switch in a lighting circuit?
   a) To control the brightness of the light
   b) To switch the light on and off from two locations
   c) To provide electrical grounding for the light fixture
   d) To protect against electrical surges

Answer: b) To switch the light on and off from two locations
Detailed Explanation: A three-way switch is used in a lighting circuit to control the on/off operation of a light fixture from two different locations. It allows the user to turn the light on or off from either switch position, providing convenience and flexibility. Three-way switches are commonly used in stairways, hallways, and rooms with multiple entrances.

177. Which of the following is a safety hazard associated with electrical systems?
   a) Grounding faults
   b) Insufficient power supply
   c) Inadequate insulation
   d) Excessive voltage drop

Answer: c) Inadequate insulation
Detailed Explanation: Inadequate insulation in electrical systems is a safety hazard that can result in electrical shocks, short circuits, and fires. Proper insulation is necessary to prevent current leakage, maintain the integrity of wiring, and ensure that electricity flows along the intended path. Any damaged or deteriorated insulation should be promptly repaired or replaced to mitigate the risk of electrical hazards.

178. Which of the following tools is commonly used to measure electrical voltage?
   a) Multimeter
   b) Wire stripper
   c) Conduit bender
   d) Cable cutter

Answer: a) Multimeter
Detailed Explanation: A multimeter is a versatile tool commonly used by electricians to measure electrical voltage, current, and resistance. It combines several functions into a single device, allowing electricians to test and troubleshoot electrical circuits. A multimeter typically has settings for measuring AC voltage, DC voltage, and other electrical parameters.

179. What is the purpose of a junction box cover in an electrical installation?
   a) To provide protection against physical damage
   b) To enhance the aesthetic appearance of the box
   c) To provide additional grounding
   d) To regulate the flow of electrical current

Answer: a) To provide protection against physical damage
Detailed Explanation: A junction box cover is used to provide physical protection to the electrical connections and wiring housed within the junction box. It helps prevent accidental contact with live wires, protects against physical damage, and reduces the risk of electrical shock. Junction box covers should always be securely installed to maintain the safety and integrity of the electrical installation.

180. Which of the following is the correct sequence for connecting wires with wire nuts?
   a) Twist the wires clockwise and secure with a wire nut.
   b) Twist the wires counterclockwise and secure with a wire nut.
   c) Twist the wires in any direction and secure them with a wire nut.
   d) Twist the wires randomly and secure with a wire nut.

Answer: a) Twist the wires clockwise and secure with a wire nut.
Detailed Explanation: When connecting wires with wire nuts, it is important to twist the wires together in a clockwise direction. This ensures a secure and tight connection. Twisting the wires counterclockwise or in random directions may result in loose connections, which can lead to electrical hazards, such as overheating or arcing.

181. What is the purpose of a conduit in electrical installations?
   a) To provide insulation for electrical wires
   b) To protect electrical wiring from physical damage
   c) To regulate electrical voltage levels
   d) To enhance the conductivity of electrical wires

Answer: b) To protect electrical wiring from physical damage
Detailed Explanation: Conduit is a protective tubing or piping system used to enclose and safeguard electrical wiring. Its primary purpose is to protect the wires from physical damage, such as impact, moisture, chemicals, and environmental conditions. Conduit also facilitates the installation, maintenance, and replacement of electrical wiring, ensuring the overall safety and longevity of the electrical system.

182. Which of the following is the most appropriate type of outlet for outdoor electrical installations?
   a) GFCI outlet

b) AFCI outlet
c) Duplex outlet
d) Isolated ground outlet

Answer: a) GFCI outlet

Detailed Explanation: Outdoor electrical installations require special precautions to ensure safety. The most appropriate type of outlet for outdoor use is a ground fault circuit interrupter (GFCI) outlet. GFCI outlets are designed to quickly detect ground faults and shut off power to prevent electrical shock. They provide enhanced protection against electrical hazards in wet or damp conditions, making them suitable for outdoor installations.

183.What is the purpose of a wire stripper in electrical work?
a) To secure wires together
b) To measure electrical current
c) To cut electrical wires
d) To remove insulation from wires

Answer: d) To remove insulation from wires

Detailed Explanation: A wire stripper is a tool used to remove the insulation from electrical wires. It has notched blades or jaws that allow precise stripping of the insulation without damaging the underlying wire. Wire strippers are essential in electrical work to ensure proper connections and to expose the conductive part of the wire for termination or splicing.

184.Which of the following is the most appropriate type of lighting for hazardous locations where flammable gasses may be present?
a) Incandescent lighting
b) Halogen lighting
c) Fluorescent lighting
d) Explosion-proof lighting

Answer: d) Explosion-proof lighting

Detailed Explanation: In hazardous locations where flammable gasses or vapors may be present, explosion-proof lighting is the most appropriate choice. Explosion-proof lighting fixtures are specially designed and constructed to prevent the ignition of flammable substances in the surrounding atmosphere. They are designed to contain any sparks, arcs, or heat generated by the lighting fixture, reducing the risk of explosions or fires in such environments.

185.Which of the following is a characteristic of a series circuit?
a) All components share the same voltage.
b) Components are connected in parallel to each other.
c) Each component has a separate path for current flow.
d) Voltage across each component varies.

Answer: a) All components share the same voltage.

Detailed Explanation: In a series circuit, all components are connected in a single path, and the same current passes through each component. As a result, all components in a series circuit share the same voltage. The total voltage of the circuit is divided among the various components.

186.What is the purpose of a busbar in an electrical panel?
a) To provide a grounding connection
b) To measure electrical current
c) To distribute electrical power
d) To protect against electrical surges

Answer: c) To distribute electrical power

Detailed Explanation: A busbar is a metal strip or bar used to distribute electrical power in an electrical panel or distribution board. It serves as a common connection point for multiple circuit breakers or fuses, allowing electrical power to be distributed to various branch circuits. The busbar ensures that electrical power is efficiently and evenly distributed throughout the electrical system.

187.Which of the following is the correct color coding for a neutral wire in electrical installations?
a) Black
b) Green
c) White
d) Red

Answer: c) White

Detailed Explanation: In electrical installations, the neutral wire is typically color-coded white. The neutral wire carries the return current from the load back to the power source, completing the electrical circuit. Color coding helps to distinguish the neutral wire from other wires in the system and aids in proper identification during installation, maintenance, and troubleshooting.

188.What is the purpose of an AFCI (Arc-Fault Circuit Interrupter) device?

a) To protect against overvoltage
b) To prevent electrical shock
c) To detect and mitigate arc faults
d) To regulate current flow

Answer: c) To detect and mitigate arc faults

Detailed Explanation: An AFCI (Arc-Fault Circuit Interrupter) device is designed to detect and mitigate arc faults in electrical circuits. Arc faults occur when there is an unintended flow of electrical current through an air gap, potentially leading to fires. AFCI devices constantly monitor the circuit for abnormal arcing conditions and quickly interrupt the circuit if an arc fault is detected, helping to prevent electrical fires.

189.Which of the following is the correct order of colors for a three-phase electrical system in the United States?

a) Black, Red, Blue
b) Red, White, Blue
c) Brown, Orange, Yellow
d) Black, White, Green

Answer: a) Black, Red, Blue

Detailed Explanation: In the United States, the correct order of colors for a three-phase electrical system is black, red, and blue. These colors are used to designate the three phases of the electrical system, with each phase having a different voltage and waveform. The specific color sequence may vary in different countries, so it is important to adhere to the local electrical standards and regulations.

190.What is the purpose of a disconnect switch in an electrical system?

a) To control the flow of current
b) To provide electrical grounding
c) To measure electrical voltage
d) To isolate electrical equipment

Answer: d) To isolate electrical equipment

Detailed Explanation: A disconnect switch is used to isolate electrical equipment from its power source. It provides a means of safely disconnecting the power supply to a specific piece of equipment or a section of the electrical system for maintenance, repairs, or in case of emergencies. The disconnect switch ensures that the equipment is de-energized and poses no electrical hazards during service or maintenance activities.

191.Which of the following is an example of a renewable energy source used in electrical generation?

a) Natural gas
b) Coal
c) Nuclear power
d) Solar power

Answer: d) Solar power

Detailed Explanation: Solar power is an example of a renewable energy source used in electrical generation. It harnesses the energy from sunlight and converts it into electricity using solar panels or photovoltaic cells. Solar power is considered renewable because it relies on an abundant and inexhaustible source—the sun—and does not deplete natural resources or release harmful emissions during operation.

192.What is the purpose of a surge protector in an electrical system?

a) To regulate voltage levels
b) To measure electrical resistance
c) To protect against power surges
d) To provide electrical grounding

Answer: c) To protect against power surges

Detailed Explanation: A surge protector, also known as a surge suppressor, is designed to protect electrical devices and equipment from power surges. Power surges can occur due to lightning strikes, voltage spikes, or sudden changes in the electrical grid. Surge protectors detect excessive voltage levels and divert the excess energy away from the connected devices, preventing damage to them. They play a vital role in safeguarding sensitive electronic equipment and preventing potential fire hazards.

193.In a residential electrical system, which wire color is typically used for the protective ground wire?
a) Black
b) Green
c) White
d) Red
Answer: b) Green
Detailed Explanation: In a residential electrical system, the protective ground wire is typically color-coded green. The ground wire provides a path for electrical current to safely flow into the ground in the event of a fault or electrical surge. It helps protect individuals from electric shock and ensures the integrity of the electrical system.
194.What does the term "load" refer to in an electrical system?
a) The electrical current flowing through a circuit
b) The resistance encountered by the electrical current
c) The power consumed by electrical devices
d) The voltage supplied by the electrical source
Answer: c) The power consumed by electrical devices
Detailed Explanation: In an electrical system, the term "load" refers to the power consumed by electrical devices or equipment connected to the circuit. It represents the energy required by the devices to perform their intended functions. The load determines the current flow, voltage drop, and overall electrical demand of the system.
195.Which of the following is the correct formula to calculate electrical resistance?
a) Resistance = Voltage / Current
b) Resistance = Current / Voltage
c) Resistance = Power / Current
d) Resistance = Voltage x Current
Answer: a) Resistance = Voltage / Current
Detailed Explanation: The correct formula to calculate electrical resistance is Resistance = Voltage / Current. Resistance is measured in ohms (Ω) and represents the opposition that a material or component offers to the flow of electrical current. Ohm's Law states that resistance is equal to the voltage across a component divided by the current flowing through it.
196.What is the purpose of a transformer in an electrical system?
a) To convert AC voltage to DC voltage
b) To regulate electrical frequency
c) To step up or step down voltage
d) To provide electrical isolation
Answer: c) To step up or step down voltage
Detailed Explanation: A transformer is an electrical device used to step up (increase) or step down (decrease) the voltage levels between two or more circuits. It operates on the principle of electromagnetic induction and is commonly used in power distribution systems to adjust voltage levels for efficient transmission and utilization. Transformers play a crucial role in delivering electrical power at appropriate voltages to different areas and loads.
197.Which of the following is a characteristic of a parallel circuit?
a) All components share the same voltage.
b) Components are connected in series to each other.
c) Each component has a separate path for current flow.
d) Voltage across each component varies.
Answer: c) Each component has a separate path for current flow.
Detailed Explanation: In a parallel circuit, each component has a separate path for current flow. This means that the current is divided among the components, and each component receives the same voltage. Unlike series circuits, where components are connected in a single path, parallel circuits allow for independent operation of each component.
198.What is the purpose of a junction box in an electrical system?
a) To support light fixtures
b) To connect electrical wires
c) To protect against electrical surges
d) To provide grounding for circuits
Answer: b) To connect electrical wires
Detailed Explanation: A junction box is used to securely connect and protect electrical wires in an electrical system. It provides a safe and enclosed space where wires can be joined together using wire nuts or other approved connectors.

Junction boxes also help prevent accidental contact with live wires and contribute to the overall safety of the electrical installation.

199. Which of the following is the correct color coding for a hot (live) wire in electrical installations?

a) Black
b) Green
c) White
d) Blue

Answer: a) Black
Detailed Explanation: In electrical installations, the hot (live) wire is typically color-coded black. The hot wire carries the current from the power source to the load, providing the energy needed for the operation of electrical devices. Color coding helps to distinguish the hot wire from other wires in the system and aids in proper identification during installation, maintenance, and troubleshooting.

200. Which of the following is a safety precaution when working with electrical equipment?

a) Wearing loose clothing and jewelry
b) Using damaged cords or plugs
c) Working on live circuits without proper training
d) Using insulated tools and equipment

Answer: d) Using insulated tools and equipment
Detailed Explanation: When working with electrical equipment, it is essential to use insulated tools and equipment to minimize the risk of electric shock. Insulated tools have handles or coatings made from non-conductive materials, such as rubber or plastic, which provide a barrier between the user and any live electrical parts. Insulated tools help prevent accidental contact with energized components, reducing the likelihood of electrical accidents or injuries.

201. What is the purpose of a junction box cover in an electrical installation?

a) To provide protection against physical damage
b) To enhance the aesthetic appearance of the box
c) To provide additional grounding
d) To regulate the flow of electrical current

Answer: a) To provide protection against physical damage
Detailed Explanation: A junction box cover is used to provide physical protection to the electrical connections and wiring housed within the junction box. It helps prevent accidental contact with live wires, protects against physical damage, and reduces the risk of electrical shock. Junction box covers should always be securely installed to maintain the safety and integrity of the electrical installation.

202. When working on an electrical installation, why is it important to turn off the power at the circuit breaker or fuse box?

a) To save energy
b) To prevent electrical surges
c) To eliminate the risk of electric shock
d) To regulate the flow of electrical current

Answer: c) To eliminate the risk of electric shock
Detailed Explanation: Turning off the power at the circuit breaker or fuse box before working on an electrical installation is crucial to eliminate the risk of electric shock. By de-energizing the circuit, you prevent the flow of electrical current through the wires, reducing the likelihood of accidental contact with live parts. This precaution ensures the safety of the person performing the work and minimizes the potential for electrical accidents or injuries.

203. What is the purpose of a ground fault circuit interrupter (GFCI) outlet?

a) To protect against overvoltage
b) To prevent electrical shock
c) To regulate current flow
d) To detect short circuits

Answer: b) To prevent electrical shock
Detailed Explanation: A ground fault circuit interrupter (GFCI) outlet is designed to detect even small imbalances in electrical current and quickly shut off power to prevent electric shock. It provides protection against ground faults, which occur when the electrical current deviates from its intended path and flows through an unintended conductor, such as a person or water. GFCI outlets are commonly used in areas where water is present, such as bathrooms, kitchens, and outdoor locations.

204.Which of the following is a unit of electrical power?

a) Ampere (A)
b) Volt (V)
c) Ohm (Ω)
d) Watt (W)

Answer: d) Watt (W)

Detailed Explanation: The unit of electrical power is the watt (W). It measures the rate of energy transfer or conversion in an electrical circuit. Power is calculated by multiplying the voltage (V) by the current (I) and is expressed in watts. The watt is the standard unit for measuring electrical power in various applications, such as appliances, lighting, and electronic devices.

205.What is the purpose of a circuit breaker in an electrical system?

a) To regulate voltage levels
b) To measure electrical resistance
c) To control the flow of current
d) To provide electrical grounding

Answer: c) To control the flow of current

Detailed Explanation: A circuit breaker is a protective device used to control and interrupt the flow of electrical current in a circuit. It is designed to automatically open the circuit and stop the flow of current when an overload or short circuit occurs. By doing so, it helps protect the electrical system and prevent damage to equipment and potential hazards such as fires.

206.Which of the following is the correct formula to calculate electrical power?

a) Power = Voltage x Current
b) Power = Current / Resistance
c) Power = Voltage / Current
d) Power = Resistance / Current

Answer: a) Power = Voltage x Current

Detailed Explanation: The formula to calculate electrical power is Power = Voltage x Current. Power is measured in watts (W) and is the product of the voltage (V) applied across a device or circuit and the current (I) flowing through it. This formula demonstrates the relationship between voltage, current, and power in an electrical system.

207.What is the purpose of a three-way switch in a lighting circuit?

a) To control the brightness of the light
b) To switch the light on and off from two locations
c) To provide electrical grounding for the light fixture
d) To protect against electrical surges

Answer: b) To switch the light on and off from two locations

Detailed Explanation: A three-way switch is used in a lighting circuit to control the on/off operation of a light fixture from two different locations. It allows the user to turn the light on or off from either switch position, providing convenience and flexibility. Three-way switches are commonly used in stairways, hallways, and rooms with multiple entrances.

208.Which of the following is a safety hazard associated with electrical systems?

a) Grounding faults
b) Insufficient power supply
c) Inadequate insulation
d) Excessive voltage drop

Answer: c) Inadequate insulation

Detailed Explanation: Inadequate insulation in electrical systems is a safety hazard that can result in electrical shocks, short circuits, and fires. Proper insulation is necessary to prevent current leakage, maintain the integrity of wiring, and ensure that electricity flows along the intended path. Any damaged or deteriorated insulation should be promptly repaired or replaced to mitigate the risk of electrical hazards.

209.Which of the following tools is commonly used to measure electrical voltage?

a) Multimeter
b) Wire stripper
c) Conduit bender
d) Cable cutter

Answer: a) Multimeter

Detailed Explanation: A multimeter is a versatile tool commonly used by electricians to measure electrical voltage, current, and resistance. It combines several functions into a single device, allowing electricians to test and troubleshoot electrical circuits. A multimeter typically has settings for measuring AC voltage, DC voltage, and other electrical parameters.

210.What is the purpose of a disconnect switch in an electrical system?

a) To control the flow of current
b) To provide electrical grounding
c) To measure electrical voltage
d) To isolate electrical equipment

Answer: d) To isolate electrical equipment

Detailed Explanation: A disconnect switch is used to isolate electrical equipment from its power source. It provides a means of safely disconnecting the power supply to a specific piece of equipment or a section of the electrical system for maintenance, repairs, or in case of emergencies. The disconnect switch ensures that the equipment is de-energized and poses no electrical hazards during service or maintenance activities.

211.In an electrical system, what is the purpose of a neutral wire?

a) To carry current from the power source to the load
b) To provide a return path for current back to the power source
c) To control the flow of current in the circuit
d) To regulate the voltage level in the system

Answer: b) To provide a return path for current back to the power source

Detailed Explanation: The neutral wire in an electrical system acts as a return path for current flowing from the load back to the power source. It completes the circuit and allows the current to flow in a closed loop. The neutral wire is typically connected to the grounded conductor at the service entrance and provides a reference point for the voltage levels in the system.

212.What is the purpose of a ground wire in an electrical system?

a) To carry current from the power source to the load
b) To provide a return path for current back to the power source
c) To protect against electrical shock and faults
d) To regulate the voltage level in the system

Answer: c) To protect against electrical shock and faults

Detailed Explanation: The ground wire in an electrical system serves as a safety measure to protect against electrical shock and faults. It provides a path for fault currents to safely flow into the ground, redirecting them away from people and equipment. The ground wire is connected to the grounding system, including grounding electrodes, to ensure proper grounding and minimize the risk of electrical hazards.

213.Which of the following is the correct formula to calculate electrical current?

a) Current = Voltage x Resistance
b) Current = Voltage / Resistance
c) Current = Power / Voltage
d) Current = Resistance / Voltage

Answer: b) Current = Voltage / Resistance

Detailed Explanation: The formula to calculate electrical current is Current = Voltage / Resistance. Electrical current is measured in amperes (A) and represents the flow of electric charge in a circuit. According to Ohm's Law, current is directly proportional to voltage and inversely proportional to resistance. Thus, by dividing the voltage by the resistance, you can determine the current flowing through a circuit.

214.What is the purpose of a junction box in an electrical system?

a) To support light fixtures
b) To connect electrical wires
c) To protect against electrical surges
d) To provide grounding for circuits

Answer: b) To connect electrical wires

Detailed Explanation: A junction box is used to securely connect electrical wires in an electrical system. It provides a safe and enclosed space where wires can be joined together using wire connectors, wire nuts, or other approved methods. Junction boxes help ensure reliable electrical connections, protect the wiring from damage, and facilitate future maintenance or modifications to the electrical system.

215.Which of the following is a safety precaution when working with electrical circuits?

a) Working with wet hands

b) Overloading circuits with excessive loads
c) Using damaged or frayed electrical cords
d) Following proper lockout/tagout procedures

Answer: d) Following proper lockout/tagout procedures

Detailed Explanation: Following proper lockout/tagout procedures is a crucial safety precaution when working with electrical circuits. Lockout/tagout procedures involve de-energizing electrical equipment, isolating it from the power source, and securing it with locks or tags to prevent accidental re-energization. This ensures the safety of workers during maintenance, repairs, or other activities involving electrical systems.

216.What is the purpose of a circuit overload protection device?
a) To regulate voltage levels
b) To measure electrical resistance
c) To control the flow of current
d) To prevent excessive current flow

Answer: d) To prevent excessive current flow

Detailed Explanation: A circuit overload protection device, such as a fuse or circuit breaker, is designed to prevent excessive current flow in a circuit. It detects when the current exceeds the rated capacity of the circuit and quickly interrupts the flow of current to prevent overheating, electrical fires, or damage to electrical equipment. The overload protection device helps maintain the safety and integrity of the electrical system.

217.Which of the following is a safety precaution when working with electrical equipment?
a) Using damaged or frayed electrical cords
b) Working alone without any supervision
c) Neglecting to use personal protective equipment
d) Ensuring proper grounding of equipment

Answer: d) Ensuring proper grounding of equipment

Detailed Explanation: Ensuring proper grounding of equipment is an important safety precaution when working with electrical equipment. Proper grounding helps prevent the buildup of electric charges and reduces the risk of electric shock. It involves connecting the equipment to a grounding system or grounding conductor, such as a grounding wire or grounding electrode, to create a safe path for electrical faults or surges.

218.What is the purpose of a transformer in an electrical system?
a) To convert AC voltage to DC voltage
b) To regulate voltage levels
c) To step up or step down voltage
d) To provide electrical isolation

Answer: c) To step up or step down voltage

Detailed Explanation: A transformer is an electrical device used to step up (increase) or step down (decrease) the voltage levels between two or more circuits. It operates on the principle of electromagnetic induction and is commonly used in power distribution systems to adjust voltage levels for efficient transmission and utilization. Transformers play a crucial role in delivering electrical power at appropriate voltages to different areas and loads.

219.Which of the following is the correct color coding for a ground wire in electrical installations?
a) Black
b) Green
c) White
d) Red

Answer: b) Green

Detailed Explanation: In electrical installations, the ground wire is typically color-coded green. The ground wire provides a safe path for fault currents to flow into the ground, protecting against electrical shock and ensuring the integrity of the electrical system. Color coding helps differentiate the ground wire from other wires and aids in proper identification during installation, maintenance, and troubleshooting.

220.What is the purpose of a junction box cover in an electrical installation?
a) To provide protection against physical damage
b) To enhance the aesthetic appearance of the box
c) To provide additional grounding
d) To regulate the flow of electrical current

Answer: a) To provide protection against physical damage
Detailed Explanation: A junction box cover is used to provide physical protection to the electrical connections and wiring housed within the junction box. It helps prevent accidental contact with live wires, protects against physical damage, and reduces the risk of electrical shock. Junction box covers should always be securely installed to maintain the safety and integrity of the electrical installation.

221.Which of the following is a characteristic of a parallel circuit?

a) All components share the same voltage.
b) Components are connected in series to each other.
c) Each component has a separate path for current flow.
d) Voltage across each component varies.

Answer: c) Each component has a separate path for current flow.
Detailed Explanation: In a parallel circuit, each component has its own separate path for current flow. This means that the current is divided among the components, and each component receives the same voltage. Unlike series circuits, where components are connected in a single path, parallel circuits allow for independent operation of each component.

222.What is the purpose of a ground fault circuit interrupter (GFCI) outlet?

a) To protect against overvoltage
b) To prevent electrical shock
c) To regulate current flow
d) To detect short circuits

Answer: b) To prevent electrical shock
Detailed Explanation: A ground fault circuit interrupter (GFCI) outlet is designed to detect even small imbalances in electrical current and quickly shut off power to prevent electric shock. It provides protection against ground faults, which occur when the electrical current deviates from its intended path and flows through an unintended conductor, such as a person or water. GFCI outlets are commonly used in areas where water is present, such as bathrooms, kitchens, and outdoor locations.

223.What is the purpose of a circuit breaker in an electrical system?

a) To regulate voltage levels
b) To measure electrical resistance
c) To control the flow of current
d) To provide electrical grounding

Answer: c) To control the flow of current
Detailed Explanation: A circuit breaker is a protective device used to control and interrupt the flow of electrical current in a circuit. It is designed to automatically open the circuit and stop the flow of current when an overload or short circuit occurs. By doing so, it helps protect the electrical system and prevent damage to equipment and potential hazards such as fires.

224.Which of the following is the correct formula to calculate electrical power?

a) Power = Voltage x Current
b) Power = Current / Resistance
c) Power = Voltage / Current
d) Power = Resistance / Current

Answer: a) Power = Voltage x Current
Detailed Explanation: The formula to calculate electrical power is Power = Voltage x Current. Power is measured in watts (W) and is the product of the voltage (V) applied across a device or circuit and the current (I) flowing through it. This formula demonstrates the relationship between voltage, current, and power in an electrical system.

225.What is the purpose of a three-way switch in a lighting circuit?

a) To control the brightness of the light
b) To switch the light on and off from two locations
c) To provide electrical grounding for the light fixture
d) To protect against electrical surges

Answer: b) To switch the light on and off from two locations
Detailed Explanation: A three-way switch is used in a lighting circuit to control the on/off operation of a light fixture from two different locations. It allows the user to turn the light on or off from either switch position, providing convenience and flexibility. Three-way switches are commonly used in stairways, hallways, and rooms with multiple entrances.

226.Which of the following is a safety hazard associated with electrical systems?

a) Grounding faults

b) Insufficient power supply
c) Inadequate insulation
d) Excessive voltage drop

Answer: c) Inadequate insulation

Detailed Explanation: Inadequate insulation in electrical systems is a safety hazard that can result in electrical shocks, short circuits, and fires. Proper insulation is necessary to prevent current leakage, maintain the integrity of wiring, and ensure that electricity flows along the intended path. Any damaged or deteriorated insulation should be promptly repaired or replaced to mitigate the risk of electrical hazards.

227.Which of the following tools is commonly used to measure electrical voltage?
a) Multimeter
b) Wire stripper
c) Conduit bender
d) Cable cutter

Answer: a) Multimeter

Detailed Explanation: A multimeter is a versatile tool commonly used by electricians to measure electrical voltage, current, and resistance. It combines several functions into a single device, allowing electricians to test and troubleshoot electrical circuits. A multimeter typically has settings for measuring AC voltage, DC voltage, and other electrical parameters.

228.What is the purpose of a disconnect switch in an electrical system?
a) To control the flow of current
b) To provide electrical grounding
c) To measure electrical voltage
d) To isolate electrical equipment

Answer: d) To isolate electrical equipment

Detailed Explanation: A disconnect switch is used to isolate electrical equipment from its power source. It provides a means of safely disconnecting the power supply to a specific piece of equipment or a section of the electrical system for maintenance, repairs, or in case of emergencies. The disconnect switch ensures that the equipment is de-energized and poses no electrical hazards during service or maintenance activities.

229.In an electrical system, what is the purpose of a neutral wire?
a) To carry current from the power source to the load
b) To provide a return path for current back to the power source
c) To control the flow of current in the circuit
d) To regulate the voltage level in the system

Answer: b) To provide a return path for current back to the power source

Detailed Explanation: The neutral wire in an electrical system acts as a return path for current flowing from the load back to the power source. It completes the circuit and allows the current to flow in a closed loop. The neutral wire is typically connected to the grounded conductor at the service entrance and provides a reference point for the voltage levels in the system.

230.What is the purpose of a ground wire in an electrical system?
a) To carry current from the power source to the load
b) To provide a return path for current back to the power source
c) To protect against electrical shock and faults
d) To regulate the voltage level in the system

Answer: c) To protect against electrical shock and faults

Detailed Explanation: The ground wire in an electrical system serves as a safety measure to protect against electrical shock and faults. It provides a path for fault currents to safely flow into the ground, redirecting them away from people and equipment. The ground wire is connected to the grounding system, including grounding electrodes, to ensure proper grounding and minimize the risk of electrical hazards.

231.Which of the following is the correct formula to calculate electrical current?
a) Current = Voltage x Resistance
b) Current = Voltage / Resistance
c) Current = Power / Voltage
d) Current = Resistance / Voltage

Answer: b) Current = Voltage / Resistance

Detailed Explanation: The formula to calculate electrical current is Current = Voltage / Resistance. Electrical current is measured in amperes (A) and represents the flow of electric charge in a circuit. According to Ohm's Law, current is directly proportional to voltage and inversely proportional to resistance. Thus, by dividing the voltage by the resistance, you can determine the current flowing through a circuit.

232.What is the purpose of a junction box in an electrical system?

a) To support light fixtures
b) To connect electrical wires
c) To protect against electrical surges
d) To provide grounding for circuits

Answer: b) To connect electrical wires

Detailed Explanation: A junction box is used to securely connect electrical wires in an electrical system. It provides a safe and enclosed space where wires can be joined together using wire connectors, wire nuts, or other approved methods. Junction boxes help ensure reliable electrical connections, protect the wiring from damage, and facilitate future maintenance or modifications to the electrical system.

233.Which of the following is a safety precaution when working with electrical circuits?

a) Working with wet hands
b) Overloading circuits with excessive loads
c) Using damaged or frayed electrical cords
d) Ensuring proper grounding of equipment

Answer: d) Ensuring proper grounding of equipment

Detailed Explanation: Ensuring proper grounding of equipment is an important safety precaution when working with electrical circuits. Proper grounding helps prevent the buildup of electric charges and reduces the risk of electric shock. It involves connecting the equipment to a grounding system or grounding conductor, such as a grounding wire or grounding electrode, to create a safe path for electrical faults or surges.

234.What is the purpose of a circuit overload protection device?

a) To regulate voltage levels
b) To measure electrical resistance
c) To control the flow of current
d) To prevent excessive current flow

Answer: d) To prevent excessive current flow

Detailed Explanation: A circuit overload protection device, such as a fuse or circuit breaker, is designed to prevent excessive current flow in a circuit. It detects when the current exceeds the rated capacity of the circuit and quickly interrupts the flow of current to prevent overheating, electrical fires, or damage to electrical equipment. The overload protection device helps maintain the safety and integrity of the electrical system.

235.Which of the following is a safety precaution when working with electrical equipment?

a) Using damaged or frayed electrical cords
b) Working alone without any supervision
c) Neglecting to use personal protective equipment
d) Ensuring proper grounding of equipment

Answer: d) Ensuring proper grounding of equipment

Detailed Explanation: Ensuring proper grounding of equipment is an important safety precaution when working with electrical equipment. Proper grounding helps prevent the buildup of electric charges and reduces the risk of electric shock. It involves connecting the equipment to a grounding system or grounding conductor, such as a grounding wire or grounding electrode, to create a safe path for electrical faults or surges.

236.What is the purpose of a disconnect switch in an electrical system?

a) To control the flow of current
b) To provide electrical grounding
c) To measure electrical voltage
d) To isolate electrical equipment

Answer: d) To isolate electrical equipment

Detailed Explanation: A disconnect switch is used to isolate electrical equipment from its power source. It provides a means of safely disconnecting the power supply to a specific piece of equipment or a section of the electrical system for maintenance, repairs, or in case of emergencies. The disconnect switch ensures that the equipment is de-energized and poses no electrical hazards during service or maintenance activities.

237.In an electrical system, what is the purpose of a neutral wire?
a) To carry current from the power source to the load
b) To provide a return path for current back to the power source
c) To control the flow of current in the circuit
d) To regulate the voltage level in the system
Answer: b) To provide a return path for current back to the power source
Detailed Explanation: The neutral wire in an electrical system acts as a return path for current flowing from the load back to the power source. It completes the circuit and allows the current to flow in a closed loop. The neutral wire is typically connected to the grounded conductor at the service entrance and provides a reference point for the voltage levels in the system.

238.What is the purpose of a ground wire in an electrical system?
a) To carry current from the power source to the load
b) To provide a return path for current back to the power source
c) To protect against electrical shock and faults
d) To regulate the voltage level in the system
Answer: c) To protect against electrical shock and faults
Detailed Explanation: The ground wire in an electrical system serves as a safety measure to protect against electrical shock and faults. It provides a path for fault currents to safely flow into the ground, redirecting them away from people and equipment. The ground wire is connected to the grounding system, including grounding electrodes, to ensure proper grounding and minimize the risk of electrical hazards.

239.Which of the following is the correct formula to calculate electrical current?
a) Current = Voltage x Resistance
b) Current = Voltage / Resistance
c) Current = Power / Voltage
d) Current = Resistance / Voltage
Answer: b) Current = Voltage / Resistance
Detailed Explanation: The formula to calculate electrical current is Current = Voltage / Resistance. Electrical current is measured in amperes (A) and represents the flow of electric charge in a circuit. According to Ohm's Law, current is directly proportional to voltage and inversely proportional to resistance. Thus, by dividing the voltage by the resistance, you can determine the current flowing through a circuit.

240.What is the purpose of a junction box in an electrical system?
a) To support light fixtures
b) To connect electrical wires
c) To protect against electrical surges
d) To provide grounding for circuits
Answer: b) To connect electrical wires
Detailed Explanation: A junction box is used to securely connect electrical wires in an electrical system. It provides a safe and enclosed space where wires can be joined together using wire connectors, wire nuts, or other approved methods. Junction boxes help ensure reliable electrical connections, protect the wiring from damage, and facilitate future maintenance or modifications to the electrical system.

241.Which of the following is a safety precaution when working with electrical circuits?
a) Working with wet hands
b) Overloading circuits with excessive loads
c) Using damaged or frayed electrical cords
d) Ensuring proper grounding of equipment
Answer: d) Ensuring proper grounding of equipment
Detailed Explanation: Ensuring proper grounding of equipment is an important safety precaution when working with electrical circuits. Proper grounding helps prevent the buildup of electric charges and reduces the risk of electric shock. It involves connecting the equipment to a grounding system or grounding conductor, such as a grounding wire or grounding electrode, to create a safe path for electrical faults or surges.

242.What is the purpose of a circuit overload protection device?
a) To regulate voltage levels
b) To measure electrical resistance

c) To control the flow of current
d) To prevent excessive current flow

Answer: d) To prevent excessive current flow

Detailed Explanation: A circuit overload protection device, such as a fuse or circuit breaker, is designed to prevent excessive current flow in a circuit. It detects when the current exceeds the rated capacity of the circuit and quickly interrupts the flow of current to prevent overheating, electrical fires, or damage to electrical equipment. The overload protection device helps maintain the safety and integrity of the electrical system.

243. Which of the following is a safety precaution when working with electrical equipment?
a) Using damaged or frayed electrical cords
b) Working alone without any supervision
c) Neglecting to use personal protective equipment
d) Ensuring proper grounding of equipment

Answer: d) Ensuring proper grounding of equipment

Detailed Explanation: Ensuring proper grounding of equipment is an important safety precaution when working with electrical equipment. Proper grounding helps prevent the buildup of electric charges and reduces the risk of electric shock. It involves connecting the equipment to a grounding system or grounding conductor, such as a grounding wire or grounding electrode, to create a safe path for electrical faults or surges.

244. What is the purpose of a disconnect switch in an electrical system?
a) To control the flow of current
b) To provide electrical grounding
c) To measure electrical voltage
d) To isolate electrical equipment

Answer: d) To isolate electrical equipment

Detailed Explanation: A disconnect switch is used to isolate electrical equipment from its power source. It provides a means of safely disconnecting the power supply to a specific piece of equipment or a section of the electrical system for maintenance, repairs, or in case of emergencies. The disconnect switch ensures that the equipment is de-energized and poses no electrical hazards during service or maintenance activities.

245. In an electrical system, what is the purpose of a neutral wire?
a) To carry current from the power source to the load
b) To provide a return path for current back to the power source
c) To control the flow of current in the circuit
d) To regulate the voltage level in the system

Answer: b) To provide a return path for current back to the power source

Detailed Explanation: The neutral wire in an electrical system acts as a return path for current flowing from the load back to the power source. It completes the circuit and allows the current to flow in a closed loop. The neutral wire is typically connected to the grounded conductor at the service entrance and provides a reference point for the voltage levels in the system.

246. What is the purpose of a ground wire in an electrical system?
a) To carry current from the power source to the load
b) To provide a return path for current back to the power source
c) To protect against electrical shock and faults
d) To regulate the voltage level in the system

Answer: c) To protect against electrical shock and faults

Detailed Explanation: The ground wire in an electrical system serves as a safety measure to protect against electrical shock and faults. It provides a path for fault currents to safely flow into the ground, redirecting them away from people and equipment. The ground wire is connected to the grounding system, including grounding electrodes, to ensure proper grounding and minimize the risk of electrical hazards.

247. Which of the following is the correct formula to calculate electrical current?
a) Current = Voltage x Resistance
b) Current = Voltage / Resistance
c) Current = Power / Voltage
d) Current = Resistance / Voltage

Answer: b) Current = Voltage / Resistance

Detailed Explanation: The formula to calculate electrical current is Current = Voltage / Resistance. Electrical current is measured in amperes (A) and represents the flow of electric charge in a circuit. According to Ohm's Law, current is directly proportional to voltage and inversely proportional to resistance. Thus, by dividing the voltage by the resistance, you can determine the current flowing through a circuit.

248.What is the purpose of a junction box in an electrical system?

a) To support light fixtures
b) To connect electrical wires
c) To protect against electrical surges
d) To provide grounding for circuits

Answer: b) To connect electrical wires

Detailed Explanation: A junction box is used to securely connect electrical wires in an electrical system. It provides a safe and enclosed space where wires can be joined together using wire connectors, wire nuts, or other approved methods. Junction boxes help ensure reliable electrical connections, protect the wiring from damage, and facilitate future maintenance or modifications to the electrical system.

249.Which of the following is a safety precaution when working with electrical circuits?

a) Working with wet hands
b) Overloading circuits with excessive loads
c) Using damaged or frayed electrical cords
d) Ensuring proper grounding of equipment

Answer: d) Ensuring proper grounding of equipment

Detailed Explanation: Ensuring proper grounding of equipment is an important safety precaution when working with electrical circuits. Proper grounding helps prevent the buildup of electric charges and reduces the risk of electric shock. It involves connecting the equipment to a grounding system or grounding conductor, such as a grounding wire or grounding electrode, to create a safe path for electrical faults or surges.

250.What is the purpose of a circuit overload protection device?

a) To regulate voltage levels
b) To measure electrical resistance
c) To control the flow of current
d) To prevent excessive current flow

Answer: d) To prevent excessive current flow

Detailed Explanation: A circuit overload protection device, such as a fuse or circuit breaker, is designed to prevent excessive current flow in a circuit. It detects when the current exceeds the rated capacity of the circuit and quickly interrupts the flow of current to prevent overheating, electrical fires, or damage to electrical equipment. The overload protection device helps maintain the safety and integrity of the electrical system.

251.Which of the following is a safety precaution when working with electrical equipment?

a) Using damaged or frayed electrical cords
b) Working alone without any supervision
c) Neglecting to use personal protective equipment
d) Ensuring proper grounding of equipment

Answer: d) Ensuring proper grounding of equipment

Detailed Explanation: Ensuring proper grounding of equipment is an important safety precaution when working with electrical equipment. Proper grounding helps prevent the buildup of electric charges and reduces the risk of electric shock. It involves connecting the equipment to a grounding system or grounding conductor, such as a grounding wire or grounding electrode, to create a safe path for electrical faults or surges.

One Liners

1. What is the purpose of a ground fault circuit interrupter (GFCI)?
   Answer: To protect against ground faults.
   Explanation: A GFCI continuously monitors the current flow in a circuit. If there is an imbalance between the current flowing in the hot and neutral conductors, indicating a ground fault, the GFCI quickly interrupts the circuit to prevent electrical shocks.
2. What is the maximum allowable voltage drop for branch circuits in a residential dwelling?
   Answer: 5%.
   Explanation: According to the National Electrical Code (NEC), the maximum allowable voltage drop for branch circuits in a residential dwelling is 5%. This ensures that the voltage at the load remains within an acceptable range.

3. What is the purpose of a transformer's core?
   Answer: To provide a path for magnetic flux and enhance transformer efficiency.
   Explanation: The core of a transformer is typically made of high-permeability material and provides a closed path for the magnetic flux generated by the primary winding, thereby enhancing the efficiency of the transformer.
4. What is the purpose of a disconnect switch in an electrical system?
   Answer: To manually disconnect power from a circuit or equipment.
   Explanation: A disconnect switch is used to manually interrupt or disconnect power from a circuit or equipment. It provides a means for safe maintenance, repair, or isolation of electrical equipment.
5. What is the purpose of a motor starter?
   Answer: To start and stop an electric motor and provide protection against overloads and short circuits.
   Explanation: A motor starter is an electrical device that controls the starting and stopping of an electric motor. It also provides protection against overloads and short circuits, ensuring the motor's safe and efficient operation.
6. What is the purpose of an electrical bonding jumper in a grounding system?
   Answer: To ensure electrical continuity between metal parts for effective grounding.
   Explanation: An electrical bonding jumper is used to establish electrical continuity between metal parts, such as electrical enclosures or equipment, to ensure effective grounding and minimize voltage potential differences.
7. What is the purpose of a time-delay fuse?
   Answer: To provide a delay before interrupting the circuit.
   Explanation: A time-delay fuse is designed to provide a delay before interrupting the circuit, allowing temporary overloads to pass through without causing an immediate interruption. This helps prevent nuisance trips.
8. What is the purpose of a ground-fault protection device?
   Answer: To protect against ground faults.
   Explanation: A ground-fault protection device, such as a ground-fault circuit interrupter (GFCI), is designed to quickly interrupt the circuit when it detects a ground fault, providing protection against electrical shock.
9. What is the purpose of a photovoltaic (PV) system in electrical installations?
   Answer: To generate electricity from sunlight.
   Explanation: A photovoltaic (PV) system uses solar panels to convert sunlight into electricity, providing a renewable energy source for electrical installations.
10. What is the purpose of a bonding jumper in an electrical system?
    Answer: To establish electrical continuity between metal parts and ensure effective grounding.
    Explanation: A bonding jumper is used to establish electrical continuity between metal parts, such as electrical enclosures or equipment, to ensure effective grounding and reduce the risk of electrical hazards.
11. What is the purpose of a ground fault locator tool?
    Answer: To locate the source of a ground fault.
    Explanation: A ground fault locator tool is used to identify and locate the source of a ground fault in an electrical circuit. It helps electricians pinpoint the exact location of the fault, aiding in troubleshooting and repair.
12. What is the purpose of a three-phase motor contactor?
    Answer: To control the operation of a three-phase motor.
    Explanation: A three-phase motor contactor is an electrical device used to control the operation of a three-phase motor. It enables the motor to start, stop, and switch directions, providing control and protection.
13. What is the purpose of a transformer's primary winding?
    Answer: To receive the input voltage.
    Explanation: The primary winding of a transformer is responsible for receiving the input voltage. It is connected to the power source and determines the primary side voltage.
14. What is the purpose of a current limiter in an electrical circuit?
    Answer: To protect against overcurrent conditions.
    Explanation: A current limiter is used to protect electrical circuits and equipment from excessive current. It limits the amount of current flowing through the circuit, preventing damage to components and ensuring safe operation.
15. What is the purpose of an electrical conduit in an electrical installation?
    Answer: To provide a protective pathway for electrical wires.
    Explanation: An electrical conduit is used to protect and enclose electrical wires, providing a safe and organized pathway. It safeguards the wires from physical damage, moisture, and other environmental factors.
16. What is the purpose of a power factor correction capacitor in an electrical system?
    Answer: To improve the power factor and increase energy efficiency.

Explanation: A power factor correction capacitor is used to improve the power factor of an electrical system. It reduces reactive power and brings the power factor closer to unity, resulting in increased energy efficiency.

17. What is the purpose of a ground rod in an electrical grounding system?
    Answer: To establish a safe path for fault currents to the earth.
    Explanation: A ground rod is driven into the ground to establish a low-resistance path for fault currents to safely dissipate into the earth, minimizing the risk of electric shock hazards.
18. What is the purpose of a fuse in an electrical circuit?
    Answer: To protect against overcurrent conditions.
    Explanation: A fuse is a protective device that interrupts the flow of current in an electrical circuit when an overcurrent condition occurs. It helps prevent damage to equipment and wiring by safely opening the circuit.
19. What is the purpose of a step-down transformer?
    Answer: To decrease the voltage level.
    Explanation: A step-down transformer is used to decrease the voltage level. It is commonly used in power distribution to lower the high-voltage transmission lines to levels suitable for commercial and residential use.
20. What is the purpose of a ground fault circuit interrupter (GFCI) in a bathroom?
    Answer: To provide protection against electrical shock hazards.
    Explanation: A GFCI is commonly installed in bathrooms to protect against electrical shock hazards. It monitors the current flowing through the circuit and quickly interrupts it if an imbalance or ground fault is detected.
21. What is the purpose of a motor overload relay?
    Answer: To protect motors from excessive current and overheating.
    Explanation: A motor overload relay is used to protect motors from excessive current and overheating. It monitors the motor's current and trips the circuit if it detects an overload condition, preventing damage to the motor.
22. What is the purpose of a ground fault protection device?
    Answer: To quickly interrupt the circuit in the event of a ground fault.
    Explanation: A ground fault protection device, such as a ground fault circuit interrupter (GFCI), is designed to quickly interrupt the circuit when it detects a ground fault. This provides protection against electrical shock by preventing the flow of current to a faulted path.
23. What is the purpose of an electrical bonding jumper?
    Answer: To establish electrical continuity between metal parts.
    Explanation: An electrical bonding jumper is used to establish electrical continuity between metal parts, such as electrical enclosures or equipment. It ensures equal potential between the metal parts and minimizes the risk of electrical shock hazards.
24. What is the purpose of a transformer's secondary winding?
    Answer: To provide the desired output voltage.
    Explanation: The secondary winding of a transformer is responsible for providing the desired output voltage based on the turns ratio of the transformer. It transfers the transformed voltage to the load.
25. What is the purpose of a circuit breaker in an electrical system?
    Answer: To protect against overcurrent and short circuit faults.
    Explanation: A circuit breaker is a protective device that automatically interrupts the flow of current in an electrical circuit when it detects an overcurrent or short circuit fault. It helps prevent damage to equipment and wiring by quickly disconnecting the circuit.
26. What is the purpose of a lightning arrester in an electrical system?
    Answer: To protect against voltage surges caused by lightning strikes.
    Explanation: A lightning arrester, also known as a surge arrester, is used to protect electrical systems and equipment from voltage surges caused by lightning strikes. It provides a low-resistance path for the surge current to safely dissipate to the ground.
27. What is the purpose of a power distribution panel in an electrical system?
    Answer: To distribute electrical power to various circuits and loads.
    Explanation: A power distribution panel is responsible for receiving electrical power from the main source and distributing it to various circuits and loads throughout a building or facility. It contains circuit breakers or fuses to protect each circuit.
28. What is the purpose of a step-up transformer?
    Answer: To increase the voltage level.

Explanation: A step-up transformer is used to increase the voltage level. It has more turns in the secondary winding than the primary winding, resulting in a higher output voltage.

29. What is the purpose of a ground fault alarm system?
    Answer: To alert when a ground fault occurs in an electrical system.
    Explanation: A ground fault alarm system is designed to detect and signal the presence of a ground fault in an electrical system. It provides an audible or visual alarm, notifying personnel of the fault so that appropriate actions can be taken.
30. What is the purpose of a bonding jumper in a grounding system?
    Answer: To provide a low-impedance path for fault current and ensure effective grounding.
    Explanation: A bonding jumper is used to establish a low-impedance path for fault current to flow safely to the ground, ensuring effective grounding of electrical systems and minimizing potential voltage differences between metal parts.
31. What is the purpose of a motor starter overload relay?
    Answer: To protect the motor against excessive current and overheating.
    Explanation: A motor starter overload relay is responsible for monitoring the current drawn by a motor. If the current exceeds a predetermined limit, indicating an overload condition, the overload relay will trip and interrupt the power supply to protect the motor from damage.
32. What is the purpose of a ground fault circuit interrupter (GFCI) receptacle?
    Answer: To provide protection against ground faults in electrical circuits.
    Explanation: A GFCI receptacle continuously monitors the current balance between the hot and neutral wires. If an imbalance occurs, indicating a ground fault, the GFCI receptacle quickly interrupts the circuit to protect against electrical shock hazards.
33. What is the purpose of a photovoltaic (PV) array combiner box in a solar power system?
    Answer: To consolidate and protect the wiring connections of multiple solar panels.
    Explanation: A PV array combiner box is used in a solar power system to combine the wiring connections from multiple solar panels into a single set of conductors. It also provides protection for the wiring against environmental factors.
34. What is the purpose of a motor control center (MCC)?
    Answer: To centralize the control and distribution of power to multiple motors.
    Explanation: A motor control center (MCC) is an assembly of motor starters, protective devices, and control equipment arranged in a centralized enclosure. It is used to control and distribute power to multiple motors in an industrial or commercial setting.
35. What is the purpose of a surge protective device (SPD) in an electrical system?
    Answer: To protect against voltage surges and transients.
    Explanation: A surge protective device (SPD) is installed in electrical systems to divert excessive voltage surges and transients safely to the ground, protecting sensitive equipment from damage and ensuring system reliability.
36. What is the purpose of a power quality analyzer in an electrical system?
    Answer: To measure and analyze electrical parameters for evaluating power quality.
    Explanation: A power quality analyzer is used to measure and analyze electrical parameters such as voltage, current, power factor, harmonics, and other factors. It helps evaluate the quality of power in an electrical system, identifying issues such as voltage sags, harmonics, and power factor problems.
37. What is the purpose of a transfer switch in a backup power system?
    Answer: To switch the power source between the utility and the backup generator during a power outage.
    Explanation: A transfer switch is used in a backup power system to automatically switch the power source between the utility and the backup generator during a power outage. It ensures a seamless transition and uninterrupted power supply to critical loads.
38. What is the purpose of a transformer's magnetic core?
    Answer: To provide a path for magnetic flux and enhance transformer efficiency.
    Explanation: The magnetic core of a transformer is made of high-permeability material and provides a closed path for the magnetic flux generated by the primary winding. It helps to enhance the efficiency of the transformer by reducing losses.
39. What is the purpose of a ground fault sensor in an electrical system?
    Answer: To sense and detect ground faults in electrical circuits.
    Explanation: A ground fault sensor is designed to sense and detect ground faults in electrical circuits. It monitors

the current flowing in the circuit and can quickly initiate protective actions, such as tripping a circuit breaker, when a ground fault is detected.

40. What is the purpose of a voltage regulator in an electrical system?
    Answer: To regulate the voltage at a constant level.
    Explanation: A voltage regulator is used to regulate the voltage in an electrical system, maintaining it at a constant level despite variations in the input voltage or load conditions. This ensures stable and reliable operation of electrical equipment and protects against voltage fluctuations.
41. What is the purpose of a ground fault locator tool?
    Answer: To locate the source of a ground fault in an electrical system.
    Explanation: A ground fault locator tool is used to identify and locate the source of a ground fault in an electrical system. It helps electricians trace the fault and pinpoint its exact location for efficient troubleshooting and repair.
42. What is the purpose of an electrical bonding jumper in a grounding system?
    Answer: To establish electrical continuity between metal parts for effective grounding and safety.
    Explanation: An electrical bonding jumper is used to connect and establish electrical continuity between metal parts, such as electrical enclosures or equipment, ensuring effective grounding and reducing the risk of electric shock hazards.
43. What is the purpose of a phase sequence meter in an electrical system?
    Answer: To determine the correct sequence of phases in a three-phase electrical system.
    Explanation: A phase sequence meter is used to identify and determine the correct sequence of phases in a three-phase electrical system. It ensures proper installation and operation of equipment that relies on the correct phase sequence.
44. What is the purpose of an AFCI (Arc-Fault Circuit Interrupter)?
    Answer: To protect against arc faults and potential fire hazards.
    Explanation: An AFCI is designed to detect and respond to arc faults, which are electrical discharges that can cause sparks and potentially start fires. It quickly interrupts the circuit to prevent fire hazards and enhance electrical safety.
45. What is the purpose of a load bank in electrical testing?
    Answer: To provide an artificial load for testing electrical systems or equipment.
    Explanation: A load bank is a device used to provide an artificial load for testing electrical systems or equipment. It allows for the assessment of performance, capacity, and efficiency by applying a controlled load to the system and measuring its response.
46. What is the purpose of a motor controller in an electrical system?
    Answer: To start, stop, and control the speed of an electric motor.
    Explanation: A motor controller is an electrical device used to start, stop, and control the speed of an electric motor by regulating the voltage or current supplied to the motor. It provides precise control and protection for motor operations.
47. What is the purpose of an electrical transformer's cooling system?
    Answer: To dissipate heat generated by the transformer.
    Explanation: An electrical transformer's cooling system is designed to dissipate the heat generated by the transformer. It ensures that the transformer remains within its temperature limits and prevents overheating, which can affect its performance and longevity.
48. What is the purpose of a shunt trip device in a circuit breaker?
    Answer: To remotely trip the circuit breaker under certain conditions.
    Explanation: A shunt trip device is an accessory used in circuit breakers to provide remote tripping capability. It allows the circuit breaker to be tripped remotely under specific conditions or in emergency situations, providing additional safety and control.
49. What is the purpose of a conductor ampacity rating in electrical installations?
    Answer: To determine the maximum current-carrying capacity of a conductor.
    Explanation: The conductor ampacity rating specifies the maximum amount of current a conductor can safely carry without exceeding its temperature rating or causing excessive voltage drop. It helps ensure that conductors are correctly sized for the intended electrical load.
50. What is the purpose of an electrical bonding bushing in a metallic conduit system?
    Answer: To provide electrical continuity and bonding between the conduit and connected equipment.
    Explanation: An electrical bonding bushing is used in a metallic conduit system to establish electrical continuity and

bonding between the conduit and connected equipment. It helps ensure effective grounding and reduces the risk of electrical shock hazards.

51. What is the purpose of a time-delay fuse?
    Answer: To provide a delay before interrupting the circuit.
    Explanation: A time-delay fuse is designed to provide a delay before interrupting the circuit, allowing temporary overloads to pass through without causing an immediate interruption. This helps prevent nuisance tripping and allows for the safe operation of equipment during short-term overloads.
52. What is the purpose of an isolating transformer?
    Answer: To electrically isolate one circuit from another.
    Explanation: An isolating transformer is used to electrically isolate one circuit from another. It provides a separation between circuits, preventing the transfer of electrical energy or potential differences and reducing the risk of electrical shock or interference.
53. What is the purpose of a motor overload protection device?
    Answer: To protect the motor against excessive current and overheating.
    Explanation: A motor overload protection device is designed to monitor the current drawn by a motor. If the current exceeds a predetermined limit or remains elevated for an extended period, indicating an overload condition, the protection device will trip and interrupt the power supply to safeguard the motor from damage.
54. What is the purpose of a ground fault sensor?
    Answer: To sense and detect ground faults in electrical circuits.
    Explanation: A ground fault sensor is used to sense and detect ground faults in electrical circuits. It monitors the current flowing in the circuit and can quickly initiate protective actions, such as tripping a circuit breaker, when a ground fault is detected. This helps prevent electrical shock hazards.
55. What is the purpose of a surge protective device (SPD) in an electrical system?
    Answer: To protect against voltage surges and transients.
    Explanation: A surge protective device (SPD) is installed in electrical systems to divert excessive voltage surges and transients safely to the ground. It helps protect sensitive equipment from damage caused by sudden voltage spikes or disturbances, ensuring system reliability and longevity.
56. What is the purpose of a motor control circuit?
    Answer: To control the operation of an electric motor.
    Explanation: A motor control circuit is responsible for controlling the operation of an electric motor by providing the necessary power, control signals, and protection features. It allows for starting, stopping, and controlling the speed and direction of the motor.
57. What is the purpose of a transformer's tap changer?
    Answer: To adjust the transformer's output voltage based on load conditions.
    Explanation: A tap changer is used in transformers to adjust the output voltage based on load conditions. It allows for fine-tuning the voltage to match the requirements of the connected equipment or compensate for variations in the input voltage.
58. What is the purpose of a motor overload relay in a motor starter?
    Answer: To protect the motor against excessive current and overheating.
    Explanation: A motor overload relay is a protective device used in motor starters to monitor the current drawn by the motor. It detects abnormal current levels indicating an overload condition and trips the circuit to prevent motor damage caused by excessive heat.
59. What is the purpose of a junction box in an electrical installation?
    Answer: To provide a safe and accessible enclosure for electrical connections.
    Explanation: A junction box is used to house and protect electrical connections, providing a safe and accessible enclosure. It helps to organize and secure wires, prevents accidental contact, and facilitates future maintenance or modifications.
60. What is the purpose of a ground fault circuit interrupter (GFCI) receptacle?
    Answer: To provide protection against ground faults and electrical shock.
    Explanation: A GFCI receptacle is designed to detect even small imbalances in current flow between the hot and neutral conductors. If a ground fault occurs, indicating current leakage, the GFCI quickly interrupts the circuit to prevent electrical shock hazards.
61. What is the purpose of a capacitor in an electrical circuit?
    Answer: To store and release electrical energy.

Explanation: A capacitor is an electrical component used to store and release electrical energy. It consists of two conductive plates separated by a dielectric material. When a voltage is applied across the plates, the capacitor stores energy in an electric field, which can be released when needed.

62. What is the purpose of a circuit breaker in an electrical system?
    Answer: To protect against overcurrent and short circuits.
    Explanation: A circuit breaker is a protective device that automatically interrupts the flow of current in an electrical circuit when it detects an overcurrent or short circuit. It serves to protect the wiring and connected equipment from damage caused by excessive current.
63. What is the purpose of a motor control center (MCC)?
    Answer: To centralize motor control and power distribution.
    Explanation: A motor control center (MCC) is an assembly of motor starters, control devices, and protective equipment housed in a single enclosure. It centralizes the control and power distribution for multiple motors, providing a convenient and organized solution.
64. What is the purpose of a ground fault locator tool?
    Answer: To locate and identify ground faults in electrical systems.
    Explanation: A ground fault locator tool is used to identify and locate ground faults in electrical systems. It helps electricians troubleshoot and pinpoint the exact location of the fault, making it easier to repair and restore system integrity.
65. What is the purpose of an electrical raceway?
    Answer: To provide a protected pathway for electrical wiring.
    Explanation: An electrical raceway, such as a conduit or cable tray, is used to provide a protected pathway for electrical wiring. It safeguards the conductors from physical damage, environmental factors, and interference, ensuring safe and organized wire routing.
66. What is the purpose of a step-down transformer?
    Answer: To decrease the voltage level.
    Explanation: A step-down transformer is designed to reduce the voltage level. It has fewer turns in the secondary winding compared to the primary winding, resulting in a lower output voltage suitable for specific applications.
67. What is the purpose of a motor overload protection device?
    Answer: To protect the motor against excessive current and overheating.
    Explanation: A motor overload protection device is used to monitor the current drawn by a motor. If the current exceeds a predetermined limit or remains elevated for an extended period, indicating an overload condition, the device will trip and interrupt the power supply to protect the motor from damage.
68. What is the purpose of a ground fault circuit interrupter (GFCI) breaker?
    Answer: To provide protection against ground faults and electrical shock.
    Explanation: A GFCI breaker works similar to a GFCI receptacle, but it is installed at the circuit breaker panel. It monitors the current flow in a circuit and quickly interrupts the circuit if it detects a ground fault, preventing electrical shock hazards.
69. What is the purpose of a current transformer (CT) in an electrical system?
    Answer: To measure current and provide a proportional reduced current output.
    Explanation: A current transformer (CT) is used to measure current in an electrical circuit. It steps down the current to a proportional reduced value that can be safely measured by instruments or protective devices, such as relays or meters.
70. What is the purpose of a surge protective device (SPD) in an electrical system?
    Answer: To protect against voltage surges and transients.
    Explanation: A surge protective device (SPD) is installed in electrical systems to divert excessive voltage surges and transients safely to the ground. It helps protect sensitive equipment from damage caused by sudden voltage spikes or disturbances, ensuring system reliability and longevity.
71. What is the purpose of an electrical bonding bushing in a metallic conduit system?
    Answer: To establish electrical continuity and bonding between the conduit and connected equipment.
    Explanation: An electrical bonding bushing is used in a metallic conduit system to ensure electrical continuity and bonding between the conduit and connected equipment. It helps maintain equipotential bonding, reducing the risk of electrical shock hazards.
72. What is the purpose of a motor starter's overload relay?
    Answer: To protect the motor against excessive current and prevent overheating.

Explanation: The overload relay in a motor starter is responsible for monitoring the current drawn by the motor. If the current exceeds a predetermined limit or remains elevated for an extended period, indicating an overload condition, the relay will trip and interrupt the power supply to safeguard the motor from damage.

73. What is the purpose of a circuit breaker's trip curve?
    Answer: To define the response time of the circuit breaker under different levels of overcurrent.
    Explanation: A circuit breaker's trip curve represents its response time to different levels of overcurrent. It helps determine the coordination and selectivity of circuit breakers in a distribution system, ensuring that the appropriate breaker trips first to isolate the fault while minimizing disruption to unaffected circuits.
74. What is the purpose of an electrical bonding grid in a facility?
    Answer: To provide equipotential bonding and minimize voltage potential differences.
    Explanation: An electrical bonding grid, also known as an equipotential bonding system, consists of conductors that interconnect metallic objects and structures within a facility. Its purpose is to establish a common reference voltage level, minimizing voltage potential differences and reducing the risk of electrical shock hazards.
75. What is the purpose of a surge protective device (SPD) in a lightning protection system?
    Answer: To divert lightning-induced surges safely to the ground.
    Explanation: A surge protective device (SPD) is an essential component of a lightning protection system. It provides a low-impedance path to safely divert lightning-induced surges to the ground, protecting the structure and electrical systems from damage caused by lightning strikes.
76. What is the purpose of an electrical disconnect switch?
    Answer: To manually interrupt the flow of electrical power to a circuit or equipment.
    Explanation: An electrical disconnect switch is used to manually interrupt the flow of electrical power to a circuit or equipment. It provides a means for safe maintenance, repair, or isolation of electrical components.
77. What is the purpose of a power factor correction capacitor bank?
    Answer: To improve the power factor and increase energy efficiency in an electrical system.
    Explanation: A power factor correction capacitor bank is used to improve the power factor of an electrical system. By introducing capacitance, it reduces reactive power and brings the power factor closer to unity, resulting in increased energy efficiency.
78. What is the purpose of a flexible metal conduit (FMC)?
    Answer: To provide a flexible, yet protective, pathway for electrical wiring.
    Explanation: Flexible metal conduit (FMC), also known as "Greenfield," is used to provide a flexible pathway for electrical wiring. It offers protection against physical damage while allowing for ease of installation in areas that require flexibility.
79. What is the purpose of a motor control circuit diagram?
    Answer: To illustrate the control and operation of a motorized system.
    Explanation: A motor control circuit diagram provides a graphical representation of the control and operation of a motorized system. It shows the connections, components, and interlocks necessary for proper motor control and protection.
80. What is the purpose of a ground electrode system in an electrical installation?
    Answer: To ensure effective grounding and the dissipation of fault currents.
    Explanation: A ground electrode system, such as grounding rods or plates, is used to provide an effective grounding connection to the earth. It ensures the dissipation of fault currents and provides a safe path for electrical energy to flow in the event of a fault.
81. What is the purpose of a voltage divider circuit?
    Answer: To obtain a desired voltage level from a higher voltage source.
    Explanation: A voltage divider circuit is used to obtain a desired voltage level from a higher voltage source. It consists of resistors connected in series or parallel to divide the voltage proportionally based on the resistor values.
82. What is the purpose of a ground fault current interrupter (GFCI) breaker?
    Answer: To provide protection against ground faults and electrical shock hazards.
    Explanation: A GFCI breaker combines the functions of a circuit breaker and a ground fault protection device. It monitors the current flowing through the circuit and quickly interrupts the power if it detects a ground fault, providing protection against electrical shock hazards.
83. What is the purpose of a capacitor start motor?
    Answer: To provide high starting torque for a single-phase induction motor.
    Explanation: A capacitor start motor is designed to provide high starting torque for single-phase induction motors.

It uses a start capacitor in conjunction with an auxiliary winding to create a phase shift and generate the necessary torque to start the motor.

84. What is the purpose of a lockout/tagout (LOTO) procedure in electrical work?
    Answer: To ensure the safe de-energization and isolation of electrical equipment during maintenance or repair.
    Explanation: A lockout/tagout (LOTO) procedure is used to ensure the safe de-energization and isolation of electrical equipment during maintenance or repair. It involves the use of locks and tags to prevent the accidental or unauthorized energization of equipment, protecting workers from electrical hazards.
85. What is the purpose of a transformer's conservator tank?
    Answer: To compensate for the expansion and contraction of transformer oil due to temperature changes.
    Explanation: A conservator tank is a container attached to a transformer that allows for the expansion and contraction of transformer oil due to temperature changes. It helps maintain the oil level and prevents moisture ingress, ensuring proper insulation and cooling of the transformer.
86. What is the purpose of a motor overload heater in a motor starter?
    Answer: To provide thermal protection for the motor against excessive current and overheating.
    Explanation: A motor overload heater is a component installed in a motor starter to provide thermal protection. It measures the current flowing through the motor and, based on its rating, heats up to trip the motor starter and protect the motor from excessive current and overheating.
87. What is the purpose of a step-voltage regulator (SVR) in a power distribution system?
    Answer: To regulate and control voltage levels within specified limits.
    Explanation: A step-voltage regulator (SVR) is used in power distribution systems to regulate and control voltage levels. It automatically adjusts the transformer taps to maintain voltage within specified limits, compensating for fluctuations and ensuring consistent voltage supply to consumers.
88. What is the purpose of a ground grid in an electrical substation?
    Answer: To provide an equipotential grounding system and safely dissipate fault currents.
    Explanation: A ground grid, consisting of interconnected conductors buried in the ground, is installed in an electrical substation to provide an equipotential grounding system. It helps safely dissipate fault currents, minimize step and touch potentials, and maintain electrical safety within the substation.
89. What is the purpose of a step-voltage regulator (SVR) in a power distribution system?
    Answer: To regulate and control voltage levels within specified limits.
    Explanation: A step-voltage regulator (SVR) is used in power distribution systems to regulate and control voltage levels. It automatically adjusts the transformer taps to maintain voltage within specified limits, compensating for fluctuations and ensuring consistent voltage supply to consumers.
90. What is the purpose of a ground grid in an electrical substation?
    Answer: To provide an equipotential grounding system and safely dissipate fault currents.
    Explanation: A ground grid, consisting of interconnected conductors buried in the ground, is installed in an electrical substation to provide an equipotential grounding system. It helps safely dissipate fault currents, minimize step and touch potentials, and maintain electrical safety within the substation.
91. What is the purpose of a power factor correction capacitor bank in an electrical system?
    Answer: To improve the power factor and reduce reactive power consumption.
    Explanation: A power factor correction capacitor bank is used to improve the power factor of an electrical system. By introducing capacitance, it compensates for the reactive power, reduces voltage drop, and improves the system's efficiency.
92. What is the purpose of a surge protective device (SPD) in an electrical system?
    Answer: To protect against voltage surges and transient overvoltages.
    Explanation: A surge protective device (SPD) is installed in electrical systems to divert excessive voltage surges and transients safely to the ground. It protects sensitive equipment from damage caused by lightning strikes, switching operations, or other events that can cause voltage spikes.
93. What is the purpose of a motor control center (MCC) bucket?
    Answer: To house motor control components, such as contactors and overload relays.
    Explanation: An MCC bucket is a housing unit within a motor control center that contains motor control components, such as contactors, overload relays, and control circuitry. It provides a compact and organized solution for controlling and protecting motors.
94. What is the purpose of an electrical interlock in a switchgear assembly?
    Answer: To prevent unsafe operations and ensure proper sequencing of switchgear operations.

Explanation: An electrical interlock is used in switchgear assemblies to prevent unsafe operations and ensure the proper sequencing of switchgear operations. It ensures that certain operations can only be performed when specific conditions are met, enhancing electrical safety.

95. What is the purpose of a phase sequence meter in an electrical system?
    Answer: To determine the correct sequence of phases in a three-phase electrical system.
    Explanation: A phase sequence meter is used to determine the correct sequence of phases in a three-phase electrical system. It ensures that the phases are correctly connected and helps avoid phase reversal issues that can cause motor and equipment damage.
96. What is the purpose of a power quality analyzer in an electrical system?
    Answer: To measure and analyze electrical parameters to assess the quality of power.
    Explanation: A power quality analyzer is used to measure and analyze electrical parameters such as voltage, current, power factor, harmonics, and other factors. It helps assess the quality of power in an electrical system, identifying issues such as voltage fluctuations, harmonics, and power factor problems.
97. What is the purpose of a line reactor in an electrical system?
    Answer: To reduce harmonics and protect equipment from voltage spikes.
    Explanation: A line reactor is used to reduce harmonics and protect electrical equipment from voltage spikes. It provides impedance to the line, smoothing out voltage variations and reducing the impact of harmonics on the system, ensuring stable and reliable operation.
98. What is the purpose of an electrical bonding jumper in a grounding system?
    Answer: To establish electrical continuity between metal parts for effective grounding and safety.
    Explanation: An electrical bonding jumper is used to establish electrical continuity between metal parts, such as electrical enclosures or equipment, ensuring effective grounding and reducing the risk of electrical shock hazards.
99. What is the purpose of a transfer switch in a backup power system?
    Answer: To automatically switch between the utility power and backup power during a power outage.
    Explanation: A transfer switch is used in a backup power system to automatically switch the power source between the utility power and backup power source (such as a generator) during a power outage. It ensures a seamless transition and uninterrupted power supply to critical loads.
100. What is the purpose of an electrical disconnect switch?
    Answer: To provide a means for manually interrupting the flow of electrical power.
    Explanation: An electrical disconnect switch is used to manually interrupt the flow of electrical power. It provides a means for safe maintenance, repair, or isolation of electrical components or equipment.
101. What is the purpose of a harmonic filter in an electrical system?
    Answer: To mitigate harmonics and improve power quality.
    Explanation: A harmonic filter is used to mitigate harmonics generated by nonlinear loads in an electrical system. It reduces harmonic distortion, improves power quality, and prevents adverse effects on equipment and power distribution systems.
102. What is the purpose of a motor control center (MCC)?
    Answer: To centralize motor control and power distribution.
    Explanation: A motor control center (MCC) is an assembly of motor starters, control devices, and protective equipment housed in a single enclosure. It centralizes motor control and power distribution, providing a convenient and organized solution for managing multiple motors.
103. What is the purpose of a surge protective device (SPD) in a data center?
    Answer: To protect sensitive electronic equipment from voltage surges and transients.
    Explanation: In a data center, a surge protective device (SPD) is installed to protect sensitive electronic equipment, such as servers and networking devices, from voltage surges and transients. It diverts excessive voltage safely to the ground, safeguarding the equipment from damage.
104. What is the purpose of a ground fault circuit interrupter (GFCI) receptacle?
    Answer: To provide protection against ground faults and electrical shock hazards.
    Explanation: A GFCI receptacle monitors the current flowing through a circuit. If it detects an imbalance or ground fault, it quickly interrupts the circuit to protect against electrical shock hazards, making it an essential safety device in areas where water is present.
105. What is the purpose of a power transformer in an electrical system?
    Answer: To step up or step down voltage for efficient transmission and distribution of electrical power.
    Explanation: A power transformer is used to step up or step down voltage levels for efficient transmission and

distribution of electrical power. It allows for long-distance transmission at high voltages and provides suitable voltages for various loads.

106. What is the purpose of a ground fault current interrupter (GFCI) receptacle?
    Answer: To quickly interrupt the flow of current in the event of a ground fault, preventing electrical shock hazards.
    Explanation: A GFCI receptacle constantly monitors the current flowing through its circuit. If an imbalance is detected, indicating a ground fault, the GFCI receptacle quickly interrupts the flow of current, providing protection against electrical shock hazards.
107. What is the purpose of a power factor correction capacitor bank in an electrical system?
    Answer: To improve the power factor and reduce reactive power consumption.
    Explanation: A power factor correction capacitor bank is used to improve the power factor of an electrical system. By compensating for reactive power, it reduces voltage drop, improves system efficiency, and reduces energy losses.
108. What is the purpose of a step-up transformer in a power distribution system?
    Answer: To increase the voltage level for efficient long-distance power transmission.
    Explanation: A step-up transformer is used to increase the voltage level for efficient long-distance power transmission. By stepping up the voltage, the current is reduced, minimizing resistive losses during transmission.
109. What is the purpose of a motor control center (MCC)?
    Answer: To house motor starters and control and protect multiple motors in an industrial setting.
    Explanation: A motor control center (MCC) is an assembly that houses motor starters, protective devices, and control circuits. It provides centralized control, protection, and distribution for multiple motors in an industrial setting.
110. What is the purpose of a surge protective device (SPD) in an electrical system?
    Answer: To divert excessive voltage surges and transients to the ground, protecting equipment from damage.
    Explanation: A surge protective device (SPD) is installed in electrical systems to protect equipment from voltage surges and transients caused by lightning strikes, switching operations, or other events. It provides a low-impedance path to safely divert excessive voltage to the ground.
111. What is the purpose of an electrical bonding conductor in a grounding system?
    Answer: To provide a low-impedance path for fault currents and ensure effective grounding.
    Explanation: An electrical bonding conductor is used to establish a low-impedance path for fault currents, ensuring effective grounding of electrical systems. It helps to protect against electrical shock hazards and ensures the proper functioning of protective devices.
112. What is the purpose of a variable frequency drive (VFD) in motor control applications?
    Answer: To control the speed and torque of an AC motor.
    Explanation: A variable frequency drive (VFD) is used to control the speed and torque of an AC motor by adjusting the frequency and voltage supplied to the motor. It allows for precise control and energy savings in motor-driven applications.
113. What is the purpose of a ground fault locator tool?
    Answer: To identify and locate ground faults in electrical systems.
    Explanation: A ground fault locator tool is used to detect and locate ground faults in electrical systems. It helps electricians identify the location of the fault and troubleshoot the system effectively for repairs and safety.
114. What is the purpose of a resonance circuit in an electrical system?
    Answer: To create a specific frequency response or filter out unwanted frequencies.
    Explanation: A resonance circuit is used to create a specific frequency response or filter out unwanted frequencies in an electrical system. It is commonly used in applications such as tuning circuits, filters, oscillators, and antennas.
115. What is the purpose of an electrical isolator (disconnector) in an electrical system?
    Answer: To provide isolation and disconnection of electrical circuits or equipment.
    Explanation: An electrical isolator, also known as a disconnector or disconnect switch, is used to provide isolation and disconnection of electrical circuits or equipment. It allows for safe maintenance, repair, or isolation of electrical components without energization.
116. What is the purpose of a load bank in electrical testing?
    Answer: To apply an artificial load to a power source for testing and commissioning purposes.
    Explanation: A load bank is used to apply an artificial load to a power source, such as a generator or UPS, to simulate real-world operating conditions during testing and commissioning. It helps verify the performance and reliability of the power source.

117. What is the purpose of a power factor correction capacitor bank in an electrical system?
Answer: To improve the power factor and reduce reactive power consumption.
Explanation: A power factor correction capacitor bank is used to improve the power factor of an electrical system by reducing reactive power. It helps reduce energy losses, improve efficiency, and minimize penalties associated with poor power factor.

118. What is the purpose of an electrical shunt trip in a circuit breaker?
Answer: To remotely trip the circuit breaker in response to a specific condition or signal.
Explanation: An electrical shunt trip is a feature in a circuit breaker that allows it to be remotely tripped using an external signal or condition. It is often used for safety or emergency shutdown purposes.

119. What is the purpose of a harmonic filter in an electrical system?
Answer: To reduce harmonic distortion and improve power quality.
Explanation: A harmonic filter is used to mitigate harmonic distortion caused by nonlinear loads in an electrical system. It helps suppress harmonics by providing a low-impedance path for the harmonic currents, improving power quality and preventing damage to sensitive equipment.

120. What is the purpose of a ground fault circuit interrupter (GFCI) in a swimming pool area?
Answer: To provide protection against electrical shock hazards in a wet environment.
Explanation: A ground fault circuit interrupter (GFCI) is essential in a swimming pool area because it provides protection against electrical shock hazards in a wet environment. It monitors the current flowing through the circuit and quickly interrupts it if a ground fault is detected, preventing electrical shocks.

121. What is the purpose of an electrical disconnect switch?
Answer: To provide a means for manually interrupting the flow of electrical power to a circuit or equipment.
Explanation: An electrical disconnect switch is used to manually interrupt the flow of electrical power to a circuit or equipment. It provides a safety mechanism for maintenance, repairs, or emergency situations by isolating the electrical supply.

122. What is the purpose of a power transformer in an electrical system?
Answer: To step up or step down voltage levels for efficient transmission and distribution of electrical power.
Explanation: A power transformer is used to either step up or step down voltage levels in an electrical system. It allows for efficient transmission and distribution of electrical power by adjusting the voltage levels as required.

123. What is the purpose of a current transformer (CT)?
Answer: To measure and monitor electrical current in a circuit.
Explanation: A current transformer (CT) is used to measure and monitor electrical current in a circuit. It steps down the current to a safe and manageable level, providing accurate readings for measurement and protection purposes.

124. What is the purpose of a power quality analyzer in an electrical system?
Answer: To measure and analyze various electrical parameters to assess the quality of power.
Explanation: A power quality analyzer is used to measure and analyze electrical parameters such as voltage, current, power factor, harmonics, and other factors in an electrical system. It helps assess the quality of power, identify issues, and ensure compliance with standards and regulations.

125. What is the purpose of an electrical bonding jumper in a grounding system?
Answer: To establish electrical continuity between metal parts for effective grounding.
Explanation: An electrical bonding jumper is used to establish electrical continuity between metal parts, such as electrical enclosures or equipment, in a grounding system. It ensures effective grounding, reduces the risk of electrical shock hazards, and equalizes potentials between metal components.

126. What is the purpose of an electrical isolator (disconnector) in a high-voltage system?
Answer: To provide a visible break and isolate electrical equipment for maintenance or safety purposes.
Explanation: An electrical isolator, also known as a disconnector or disconnect switch, is used in high-voltage systems to provide a visible break and isolate electrical equipment. It allows for safe maintenance, repairs, or emergency situations by isolating the equipment from the power source.

127. What is the purpose of a power factor correction capacitor bank in an industrial setting?
Answer: To improve the power factor and reduce reactive power consumption.
Explanation: A power factor correction capacitor bank is used in an industrial setting to improve the power factor of an electrical system. It reduces reactive power consumption, improves system efficiency, and minimizes penalties associated with poor power factor.

128. What is the purpose of a generator transfer switch?
    Answer: To safely switch the electrical load between utility power and generator power during power outages.
    Explanation: A generator transfer switch is used to safely transfer the electrical load from utility power to generator power during power outages. It ensures proper sequencing and prevents back-feeding, providing a reliable and safe backup power solution.
129. What is the purpose of a motor control center (MCC)?
    Answer: To centralize motor control and power distribution in an industrial facility.
    Explanation: A motor control center (MCC) is an assembly of motor starters, control devices, and protective equipment housed in a centralized enclosure. It provides a convenient and organized solution for motor control and power distribution in an industrial facility.
130. What is the purpose of a ground fault circuit interrupter (GFCI) receptacle?
    Answer: To provide protection against ground faults and electrical shock hazards.
    Explanation: A GFCI receptacle continuously monitors the current flowing through it. If an imbalance is detected, indicating a ground fault, it quickly interrupts the circuit to protect against electrical shock hazards caused by current leakage.
131. What is the purpose of a synchronous motor?
    Answer: To convert electrical energy into mechanical energy at a fixed speed.
    Explanation: A synchronous motor is designed to convert electrical energy into mechanical energy at a fixed speed that is synchronous with the frequency of the applied voltage. It is often used in applications where precise speed control and synchronization are required.
132. What is the purpose of a lightning arrester in an electrical system?
    Answer: To protect electrical equipment from voltage surges caused by lightning strikes.
    Explanation: A lightning arrester, also known as a surge arrester or lightning diverter, is installed in electrical systems to protect equipment from voltage surges caused by lightning strikes. It provides a low-impedance path to safely discharge the excess current to the ground.
133. What is the purpose of a power transformer in a distribution system?
    Answer: To step down the voltage for local distribution to consumers.
    Explanation: A power transformer in a distribution system is used to step down the voltage from the transmission level to a lower voltage suitable for local distribution to consumers. It ensures efficient and safe delivery of electrical power to end-users.
134. What is the purpose of a distribution panelboard in an electrical system?
    Answer: To distribute electricity to various circuits and loads within a building.
    Explanation: A distribution panelboard is a component of an electrical system that receives power from the main service and distributes it to various circuits and loads within a building. It contains circuit breakers or fuses to protect each individual circuit.
135. What is the purpose of an electrical bonding bushing in a metallic conduit system?
    Answer: To provide electrical continuity and establish bonding between metallic components.
    Explanation: An electrical bonding bushing is used in a metallic conduit system to establish electrical continuity and bonding between metallic components, such as conduits, enclosures, and equipment. It helps ensure effective grounding and reduces the risk of electric shock hazards.
136. What is the purpose of a capacitor in a single-phase motor?
    Answer: To create a phase shift and facilitate the motor's starting torque.
    Explanation: In a single-phase motor, a capacitor is used to create a phase shift between the main winding and the auxiliary winding. This phase shift helps generate a rotating magnetic field and provides the necessary starting torque to initiate motor rotation.
137. What is the purpose of a rectifier in an electrical system?
    Answer: To convert alternating current (AC) to direct current (DC).
    Explanation: A rectifier is used to convert alternating current (AC) to direct current (DC) by allowing current flow in one direction only. It is commonly used in power supplies and electronic devices that require DC power for operation.
138. What is the purpose of a power factor meter in an electrical system?
    Answer: To measure and monitor the power factor of an electrical load.
    Explanation: A power factor meter is used to measure and monitor the power factor of an electrical load. It helps

assess the efficiency of power usage, identify power factor correction needs, and ensure compliance with power factor requirements.

139. What is the purpose of an electrical disconnect switch in a photovoltaic (PV) system?
Answer: To isolate the PV system from the electrical grid during maintenance or emergencies.
Explanation: An electrical disconnect switch in a PV system is used to isolate the PV system from the electrical grid. It provides a means to disconnect the PV system from the grid for maintenance, repairs, or emergency situations, ensuring the safety of personnel working on the system.

140. What is the purpose of a load shedding system in an electrical distribution network?
Answer: To prioritize and shed non-essential loads during periods of high demand or system stress.
Explanation: A load shedding system is used in an electrical distribution network to prioritize and shed non-essential loads during periods of high demand or system stress. It helps maintain the stability and balance of the electrical grid by reducing the demand to match the available supply.

141. What is the purpose of a motor control contactor in an electrical system?
Answer: To switch the power supply to a motor and control its operation.
Explanation: A motor control contactor is an electrical device used to switch the power supply to a motor and control its operation. It allows for remote control, protection, and efficient operation of the motor.

142. What is the purpose of a step-voltage regulator (SVR) in a power distribution system?
Answer: To regulate and maintain a specified voltage level within an acceptable range.
Explanation: A step-voltage regulator (SVR) is used in a power distribution system to regulate and maintain a specified voltage level within an acceptable range. It adjusts the transformer taps to compensate for voltage fluctuations and ensure consistent voltage supply to consumers.

143. What is the purpose of an electrical disconnect switch in a circuit?
Answer: To provide a means of isolating electrical equipment for maintenance or repair.
Explanation: An electrical disconnect switch is used to provide a means of isolating electrical equipment from the power source for maintenance, repair, or emergency situations. It ensures the safety of personnel working on the equipment.

144. What is the purpose of a power quality analyzer in an electrical system?
Answer: To measure and analyze various electrical parameters to assess power quality.
Explanation: A power quality analyzer is used to measure and analyze various electrical parameters such as voltage, current, power factor, harmonics, and more. It helps assess the quality of power in an electrical system, identify issues, and ensure compliance with standards and regulations.

145. What is the purpose of an electrical bonding bushing in a metallic conduit system?
Answer: To provide electrical continuity and bonding between metallic components for effective grounding.
Explanation: An electrical bonding bushing is used in a metallic conduit system to provide electrical continuity and bonding between metallic components such as conduits, enclosures, and equipment. It helps establish effective grounding and reduce the risk of electrical shock hazards.

146. What is the purpose of a power factor correction capacitor bank in an electrical system?
Answer: To improve power factor and reduce reactive power demand.
Explanation: A power factor correction capacitor bank is utilized to improve the power factor of an electrical system by reducing reactive power demand. By supplying reactive power locally, it reduces the burden on the utility grid, improves energy efficiency, and minimizes penalties associated with poor power factor.

147. What is the purpose of an electrical disconnect switch in a motor circuit?
Answer: To provide a means for isolating and de-energizing the motor for maintenance or repair.
Explanation: An electrical disconnect switch in a motor circuit is used to isolate and de-energize the motor from the power source. It allows for safe maintenance, repair, or replacement of motor components without risk of electrical shock hazards.

148. What is the purpose of a voltage regulator in an electrical system?
Answer: To regulate and stabilize the voltage at a desired level.
Explanation: A voltage regulator is used to regulate and stabilize the voltage in an electrical system, ensuring it remains within a desired range. It compensates for voltage variations caused by fluctuations in the power supply or changes in the load demand, providing consistent voltage levels to connected equipment.

149. What is the purpose of a ground fault circuit interrupter (GFCI) breaker?
Answer: To provide protection against ground faults and electrical shock hazards.
Explanation: A GFCI breaker combines the functions of a circuit breaker and a ground fault protection device. It

monitors the current flow and quickly interrupts the circuit if it detects a ground fault. This provides enhanced protection against electrical shock hazards caused by current leakage.

150. What is the purpose of a power monitoring system in an electrical installation?
    Answer: To measure, monitor, and analyze electrical parameters to optimize energy usage and diagnose issues.
    Explanation: A power monitoring system is used in an electrical installation to measure, monitor, and analyze various electrical parameters such as voltage, current, power factor, energy consumption, and more. It helps optimize energy usage, identify inefficiencies, diagnose problems, and improve overall system performance.
151. What is the purpose of an electrical bonding jumper in a grounding system?
    Answer: To establish electrical continuity and bonding between metal parts for effective grounding and safety.
    Explanation: An electrical bonding jumper is used to connect and establish electrical continuity between metal parts, such as electrical enclosures or equipment, in a grounding system. It ensures effective grounding, reduces the risk of electrical shock hazards, and equalizes potentials between metal components.
152. What is the purpose of a motor control center (MCC)?
    Answer: To centralize motor control and power distribution in an industrial setting.
    Explanation: A motor control center (MCC) is an assembly of motor starters, control devices, and protective equipment housed in a centralized enclosure. It provides a convenient and organized solution for motor control and power distribution in an industrial setting.
153. What is the purpose of a ground fault circuit interrupter (GFCI) outlet?
    Answer: To prevent electrical shock - A GFCI outlet is designed to detect even small imbalances in electrical current and quickly shut off power to prevent electric shock. It provides protection against ground faults, which occur when the electrical current deviates from its intended path and flows through an unintended conductor, such as a person or water. GFCI outlets are commonly used in areas where water is present, such as bathrooms, kitchens, and outdoor locations.
154. What is the minimum wire size allowed for a 20-ampere branch circuit?
    Answer: 12 AWG - According to the National Electrical Code (NEC), a 20-ampere branch circuit should be wired with a minimum of 12 AWG (American Wire Gauge) copper wire. This wire size is capable of handling the current without excessive voltage drop or overheating.
155. Which of the following is the standard voltage for residential electrical systems in the United States?
    Answer: 120 volts - The standard voltage for residential electrical systems in the United States is 120 volts. This voltage is used for general lighting, small appliances, and most household outlets. Higher voltages, such as 240 volts, are typically used for larger appliances and specialized equipment.
156. When installing electrical wiring in a damp location, which type of wiring method would be most appropriate?
    Answer: Liquidtight flexible metal conduit (LFMC) - In damp locations, it is important to use wiring methods that provide protection against moisture. Liquidtight flexible metal conduit (LFMC) is a suitable choice as it is designed to be resistant to liquids, including water. It provides a high level of protection against moisture and is commonly used in areas such as basements, garages, and outdoor installations.
157. Which of the following is an example of a single-pole switch?
    Answer: Toggle switch - A single-pole switch is the most basic type of switch used to control a light or electrical device from a single location. A toggle switch is a common example of a single-pole switch, where a flip of the toggle either turns the device on or off.
158. What is the purpose of a junction box in an electrical system?
    Answer: To connect electrical wires - A junction box is used to securely connect and protect electrical wires in an electrical system. It provides a safe and enclosed space where wires can be joined together using wire nuts or other approved connectors. Junction boxes also help prevent accidental contact with live wires and contribute to the overall safety of the electrical installation.
159. Which of the following is the correct formula to calculate electrical power?
    Answer: Power = Voltage x Current - The formula to calculate electrical power is Power = Voltage x Current. Power is measured in watts (W) and is the product of the voltage (V) applied across a device or circuit and the current (I) flowing through it. This formula demonstrates the relationship between voltage, current, and power in an electrical system.
160. What is the purpose of a circuit breaker in an electrical system?
    Answer: To control the flow of current - A circuit breaker is a protective device used to control and interrupt the flow of electrical current in a circuit. It is designed to automatically open the circuit and stop the flow of current

when an overload or short circuit occurs. By doing so, it helps protect the electrical system and prevent damage to equipment and potential hazards such as fires.

161. Which of the following is a unit of electrical resistance?
Answer: Ohm (Ω) - The unit of electrical resistance is the ohm (Ω). It is named after the German physicist Georg Simon Ohm and is represented by the Greek letter omega (Ω). Resistance is a measure of how much an electrical component or material opposes the flow of electric current.

162. What does the term "grounding" refer to in electrical systems?
Answer: Connecting electrical devices to the Earth - Grounding in electrical systems refers to the process of connecting electrical devices, equipment, and structures to the Earth or a grounding electrode system. This connection serves several purposes, including providing a safe path for electrical faults, reducing electrical noise, and ensuring the effectiveness of overcurrent protection devices.

163. Which of the following symbols represents a resistor in an electrical circuit diagram?
Answer: A zigzag line - In electrical circuit diagrams, a zigzag line is commonly used to represent a resistor. A resistor is an electronic component that restricts the flow of electric current, dissipating electrical energy in the form of heat. It is often used to control the amount of current or voltage in a circuit.

164. What is the purpose of a transformer in an electrical system?
Answer: To step up or step down voltage - A transformer is an electrical device used to transfer electrical energy between two or more circuits through electromagnetic induction. Its primary function is to step up (increase) or step down (decrease) the voltage levels between the input and output circuits. Transformers are commonly used in power distribution systems to adjust voltage levels for165. Which of the following is an example of a renewable energy source used in electrical generation?
Answer: Solar power - Solar power is an example of a renewable energy source used in electrical generation. It harnesses the energy from sunlight and converts it into electricity using solar panels or photovoltaic cells. Solar power is considered renewable because it relies on an abundant and inexhaustible source—the sun—and does not deplete natural resources or release harmful emissions during operation.

165. What is the purpose of a three-way switch in a lighting circuit?
Answer: To control the on/off operation of a light fixture from two different locations - A three-way switch is used in a lighting circuit to control the on/off operation of a light fixture from two different locations. It allows the user to turn the light on or off from either switch position, providing convenience and flexibility. Three-way switches are commonly used in stairways, hallways, and rooms with multiple entrances.

166. Which of the following is a safety hazard associated with electrical systems?
Answer: Inadequate insulation - Inadequate insulation in electrical systems is a safety hazard that can result in electrical shocks, short circuits, and fires. Proper insulation is necessary to prevent current leakage, maintain the integrity of wiring, and ensure that electricity flows along the intended path. Any damaged or deteriorated insulation should be promptly repaired or replaced to mitigate the risk of electrical hazards.

167. Which of the following tools is commonly used to measure electrical voltage?
Answer: Multimeter - A multimeter is a versatile tool commonly used by electricians to measure electrical voltage, current, and resistance. It combines several functions into a single device, allowing electricians to test and troubleshoot electrical circuits. A multimeter typically has settings for measuring AC voltage, DC voltage, and other electrical parameters.

168. What is the purpose of a disconnect switch in an electrical system?
Answer: To isolate electrical equipment - A disconnect switch is used to isolate electrical equipment from its power source. It provides a means of safely disconnecting the power supply to a specific piece of equipment or a section of the electrical system for maintenance, repairs, or in case of emergencies. The disconnect switch ensures that the equipment is de-energized and poses no electrical hazards during service or maintenance activities.

169. In an electrical system, what is the purpose of a neutral wire?
Answer: To provide a return path for current back to the power source - The neutral wire in an electrical system acts as a return path for current flowing from the load back to the power source. It completes the circuit and allows the current to flow in a closed loop. The neutral wire is typically connected to the grounded conductor at the service entrance and provides a reference point for the voltage levels in the system.

170. What is the purpose of a ground wire in an electrical system?
Answer: To protect against electrical shock and faults - The ground wire in an electrical system serves as a safety measure to protect against electrical shock and faults. It provides a path for fault currents to safely flow into the

ground, redirecting them away from people and equipment. The ground wire is connected to the grounding system, including grounding electrodes, to ensure proper grounding and minimize the risk of electrical hazards.

171. Which of the following is the correct formula to calculate electrical current?
    Answer: Current = Voltage / Resistance - The formula to calculate electrical current is Current = Voltage / Resistance. Electrical current is measured in amperes (A) and represents the flow of electric charge in a circuit. According to Ohm's Law, current is directly proportional to voltage and inversely proportional to resistance. Thus, by dividing the voltage by the resistance, you can determine the current flowing through a circuit.
172. What is the purpose of a junction box in an electrical system?
    Answer: To connect electrical wires - A junction box is used to securely connect and protect electrical wires in an electrical system. It provides a safe and enclosed space where wires can be joined together using wire connectors, wire nuts, or other approved methods. Junction boxes help ensure reliable electrical connections, protect the wiring from damage, and facilitate future maintenance or modifications to the electrical system.
173. Which of the following is a unit of electrical resistance?
    Answer: Ohm (Ω) - The unit of electrical resistance is the ohm (Ω). It is named after the German physicist Georg Simon Ohm and is represented by the Greek letter omega (Ω). Resistance is a measure of how much an electrical component or material opposes the flow of electric current.
174. What does the term "grounding" refer to in electrical systems?
    Answer: Connecting electrical devices to the Earth - Grounding in electrical systems refers to the process of connecting electrical devices, equipment, and structures to the Earth or a grounding electrode system. This connection serves several purposes, including providing a safe path for electrical faults, reducing electrical noise, and ensuring the effectiveness of overcurrent protection devices.
175. What is the purpose of a transformer in an electrical system?
    Answer: To step up or step down voltage - A transformer is an electrical device used to transfer electrical energy between two or more circuits through electromagnetic induction. Its primary function is to step up (increase) or step down (decrease) the voltage levels between the input and output circuits. Transformers are commonly used in power distribution systems to adjust voltage levels for efficient transmission and utilization.
176. Which of the following is an example of a renewable energy source used in electrical generation?
    Answer: Solar power - Solar power is an example of a renewable energy source used in electrical generation. It harnesses the energy from sunlight and converts it into electricity using solar panels or photovoltaic cells. Solar power is considered renewable because it relies on an abundant and inexhaustible source—the sun—and does not deplete natural resources or release harmful emissions during operation.
177. What is the purpose of a three-way switch in a lighting circuit?
    Answer: To control the on/off operation of a light fixture from two different locations - A three-way switch is used in a lighting circuit to control the on/off operation of a light fixture from two different locations. It allows the user to turn the light on or off from either switch position, providing convenience and flexibility. Three-way switches are commonly used in stairways, hallways, and rooms with multiple entrances.
178. Which of the following is a safety hazard associated with electrical systems?
    Answer: Inadequate insulation - Inadequate insulation in electrical systems is a safety hazard that can result in electrical shocks, short circuits, and fires. Proper insulation is necessary to prevent current leakage, maintain the integrity of wiring, and ensure that electricity flows along the intended path. Any damaged or deteriorated insulation should be promptly repaired or replaced to mitigate the risk of electrical hazards.
179. Which of the following tools is commonly used to measure electrical voltage?
    Answer: Multimeter - A multimeter is a versatile tool commonly used by electricians to measure electrical voltage, current, and resistance. It combines several functions into a single device, allowing electricians to test and troubleshoot electrical circuits. A multimeter typically has settings for measuring AC voltage, DC voltage, and other electrical parameters.
180. What is the purpose of a disconnect switch in an electrical system?
    Answer: To isolate electrical equipment - A disconnect switch is used to isolate electrical equipment from its power source. It provides a means of safely disconnecting the power supply to a specific piece of equipment or a section of the electrical system for maintenance, repairs, or in case of emergencies. The disconnect switch ensures that the equipment is de-energized and poses no electrical hazards during service or maintenance activities.
181. In an electrical system, what is the purpose of a neutral wire?
    Answer: To provide a return path for current back to the power source - The neutral wire in an electrical system acts as a return path for current flowing from the load back to the power source. It completes the circuit and allows

the current to flow in a closed loop. The neutral wire is typically connected to the grounded conductor at the service entrance and provides a reference point for the voltage levels in the system.

182.What is the purpose of a ground wire in an electrical system?
Answer: To protect against electrical shock and faults - The ground wire in an electrical system serves as a safety measure to protect against electrical shock and faults. It provides a path for fault currents to safely flow into the ground, redirecting them away from people and equipment. The ground wire is connected to the grounding system, including grounding electrodes, to ensure proper grounding and minimize the risk of electrical hazards.

183.Which of the following is the correct formula to calculate electrical current?
Answer: Current = Voltage / Resistance - The formula to calculate electrical current is Current = Voltage / Resistance. Electrical current is measured in amperes (A) and represents the flow of electric charge in a circuit. According to Ohm's Law, current is directly proportional to voltage and inversely proportional to resistance. Thus, by dividing the voltage by the resistance, you can determine the current flowing through a circuit.

184.What is the purpose of a junction box in an electrical system?
Answer: To connect electrical wires - A junction box is used to securely connect and protect electrical wires in an electrical system. It provides a safe and enclosed space where wires can be joined together using wire connectors, wire nuts, or other approved methods. Junction boxes help ensure reliable electrical connections, protect the wiring from damage, and facilitate future maintenance or modifications to the electrical system.

185.Which of the following is a unit of electrical resistance?
Answer: Ohm (Ω) - The unit of electrical resistance is the ohm (Ω). It is named after the German physicist Georg Simon Ohm and is represented by 1877. Which of the following is a unit of electrical resistance?
Answer: Ohm (Ω) - The unit of electrical resistance is the ohm (Ω). It is named after the German physicist Georg Simon Ohm and is represented by the Greek letter omega (Ω). Resistance is a measure of how much an electrical component or material opposes the flow of electric current.

186.What does the term "grounding" refer to in electrical systems?
Answer: Connecting electrical devices to the Earth - Grounding in electrical systems refers to the process of connecting electrical devices, equipment, and structures to the Earth or a grounding electrode system. This connection serves several purposes, including providing a safe path for electrical faults, reducing electrical noise, and ensuring the effectiveness of overcurrent protection devices.

187.Which of the following symbols represents a resistor in an electrical circuit diagram?
Answer: A zigzag line - In electrical circuit diagrams, a zigzag line is commonly used to represent a resistor. A resistor is an electronic component that restricts the flow of electric current, dissipating electrical energy in the form of heat. It is often used to control the amount of current or voltage in a circuit.

188.What is the purpose of a transformer in an electrical system?
Answer: To step up or step down voltage - A transformer is an electrical device used to transfer electrical energy between two or more circuits through electromagnetic induction. Its primary function is to step up (increase) or step down (decrease) the voltage levels between the input and output circuits. Transformers are commonly used in power distribution systems to adjust voltage levels for efficient transmission and utilization.

189.Which of the following is an example of a renewable energy source used in electrical generation?
Answer: Solar power - Solar power is an example of a renewable energy source used in electrical generation. It harnesses the energy from sunlight and converts it into electricity using solar panels or photovoltaic cells. Solar power is considered renewable because it relies on an abundant and inexhaustible source—the sun—and does not deplete natural resources or release harmful emissions during operation.

190.What is the purpose of a three-way switch in a lighting circuit?
Answer: To control the on/off operation of a light fixture from two different locations - A three-way switch is used in a lighting circuit to control the on/off operation of a light fixture from two different locations. It allows the user to turn the light on or off from either switch position, providing convenience and flexibility. Three-way switches are commonly used in stairways, hallways, and rooms with multiple entrances.

191.Which of the following is a safety hazard associated with electrical systems?
Answer: Inadequate insulation - Inadequate insulation in electrical systems is a safety hazard that can result in electrical shocks, short circuits, and fires. Proper insulation is necessary to prevent current leakage, maintain the integrity of wiring, and ensure that electricity flows along the intended path. Any damaged or deteriorated insulation should be promptly repaired or replaced to mitigate the risk of electrical hazards.

192.Which of the following tools is commonly used to measure electrical voltage?
Answer: Multimeter - A multimeter is a versatile tool commonly used by electricians to measure electrical voltage,

current, and resistance. It combines several functions into a single device, allowing electricians to test and troubleshoot electrical circuits. A multimeter typically has settings for measuring AC voltage, DC voltage, and other electrical parameters.

193. What is the purpose of a disconnect switch in an electrical system?
Answer: To isolate electrical equipment - A disconnect switch is used to isolate electrical equipment from its power source. It provides a means of safely disconnecting the power supply to a specific piece of equipment or a section of the electrical system for maintenance, repairs, or in case of emergencies. The disconnect switch ensures that the equipment is de-energized and poses no electrical hazards during service or maintenance activities.
194. In an electrical system, what is the purpose of a neutral wire?
Answer: To provide a return path for current back to the power source - The neutral wire in an electrical system acts as a return path for current flowing from the load back to the power source. It completes the circuit and allows the current to flow in a closed loop. The neutral wire is typically connected to the grounded conductor at the service entrance and provides a reference point for the voltage levels in the system.
195. What is the purpose of a ground wire in an electrical system?
Answer: To protect against electrical shock and faults - The ground wire in an electrical system serves as a safety measure to protect against electrical shock and faults. It provides a path for fault currents to safely flow into the ground, redirecting them away from people and equipment. The ground wire is connected to the grounding system, including grounding electrodes, to ensure proper grounding and minimize the risk of electrical hazards.
196. Which of the following is the correct formula to calculate electrical current?
Answer: Current = Voltage / Resistance - The formula to calculate electrical current is Current = Voltage / Resistance. Electrical current is measured in amperes (A) and represents the flow of electric charge in a circuit. According to Ohm's Law, current is directly proportional to voltage and inversely proportional to resistance. Therefore, by dividing the voltage by the resistance, you can determine the current flowing through a circuit.
197. What is the purpose of a junction box in an electrical system?
Answer: To connect electrical wires - A junction box is used to securely connect and protect electrical wires in an electrical system. It provides a safe and enclosed space where wires can be joined together using wire connectors, wire nuts, or other approved methods. Junction boxes help ensure reliable electrical connections, protect the wiring from damage, and facilitate future maintenance or modifications to the electrical system.
198. Which of the following is a unit of electrical resistance?
Answer: Ohm (Ω) - The unit of electrical resistance is the ohm (Ω). It is named after the German physicist Georg Simon Ohm and is represented by the Greek letter omega (Ω). Resistance is a measure of how much an electrical component or material opposes the flow of electric current. Higher resistance values impede current flow, while lower resistance values allow for easier current flow.
199. What does the term "grounding" refer to in electrical systems?
Answer: Connecting electrical devices to the Earth - Grounding in electrical systems refers to the process of connecting electrical devices, equipment, and structures to the Earth or a grounding electrode system. This connection serves several purposes, including providing a safe path for electrical faults, reducing electrical noise, and ensuring the effectiveness of overcurrent protection devices. Grounding helps to prevent electric shock and protects both people and equipment from electrical hazards.
200. Which of the following symbols represents a resistor in an electrical circuit diagram?
Answer: A zigzag line - In electrical circuit diagrams, a zigzag line is commonly used to represent a resistor. A resistor is an electronic component that restricts the flow of electric current, dissipating electrical energy in the form of heat. It is often used to control the amount of current or voltage in a circuit and is an essential component in many electrical applications.
201. What is the purpose of a transformer in an electrical system?
Answer: To step up or step down voltage - A transformer is an electrical device used to transfer electrical energy between two or more circuits through electromagnetic induction. Its primary function is to step up (increase) or step down (decrease) the voltage levels between the input and output circuits. Transformers are commonly used in power distribution systems to adjust voltage levels for efficient transmission and utilization. They play a crucial role in delivering electricity at appropriate voltages to various locations and devices.
202. Which of the following is an example of a renewable energy source used in electrical generation?
Answer: Wind power - Wind power is an example of a renewable energy source used in electrical generation. It harnesses the kinetic energy of wind and converts it into electricity using wind turbines. Wind power is considered

renewable because it relies on a constantly replenished source—the wind—and does not deplete natural resources or produce harmful emissions during operation.

203. What is the purpose of a three-way switch in a lighting circuit?
Answer: To control the on/off operation of a light fixture from two different locations - A three-way switch is used in a lighting circuit to control the on/off operation of a light fixture from two different locations. It allows the user to turn the light on or off from either switch position, providing convenience and flexibility. Three-way switches are commonly used in rooms with multiple entrances or where controlling the lighting from different points is desired.

204. Which of the following is a safety hazard associated with electrical systems?
Answer: Inadequate insulation - Inadequate insulation in electrical systems can pose a safety hazard. Insulation is designed to prevent the flow of electric current to unintended paths and protect against electrical shock and short circuits. If insulation is damaged, worn out, or improperly installed, it can lead to electrical faults, fire hazards, or the risk of electric shock. Regular inspection and maintenance of electrical insulation are essential to ensure a safe electrical system.

205. Which of the following tools is commonly used to measure electrical voltage?
Answer: Digital voltmeter - A digital voltmeter is a commonly used tool to measure electrical voltage. It is a precise and versatile instrument that can accurately measure both AC and DC voltage levels in electrical circuits. Digital voltmeters provide numerical readings on a digital display, making it easier to read and interpret voltage values. They are essential for troubleshooting electrical systems, verifying voltage levels, and ensuring electrical safety.

206. What is the purpose of a disconnect switch in an electrical system?
Answer: To isolate electrical equipment - A disconnect switch is used to isolate electrical equipment from its power source. It provides a means of safely disconnecting the power supply to a specific piece of equipment or a section of the electrical system209. What is the purpose of a disconnect switch in an electrical system?
Answer: To isolate electrical equipment - A disconnect switch is used to isolate electrical equipment from its power source. It provides a means of safely disconnecting the power supply to a specific piece of equipment or a section of the electrical system for maintenance, repairs, or in case of emergencies. The disconnect switch ensures that the equipment is de-energized and poses no electrical hazards during service or maintenance activities.

207. In an electrical system, what is the purpose of a neutral wire?
Answer: To provide a return path for current back to the power source - The neutral wire in an electrical system acts as a return path for current flowing from the load back to the power source. It completes the circuit and allows the current to flow in a closed loop. The neutral wire is typically connected to the grounded conductor at the service entrance and provides a reference point for the voltage levels in the system.

208. What is the purpose of a ground wire in an electrical system?
Answer: To protect against electrical shock and faults - The ground wire in an electrical system serves as a safety measure to protect against electrical shock and faults. It provides a path for fault currents to safely flow into the ground, redirecting them away from people and equipment. The ground wire is connected to the grounding system, including grounding electrodes, to ensure proper grounding and minimize the risk of electrical hazards.

209. Which of the following is the correct formula to calculate electrical current?
Answer: Current = Voltage / Resistance - The formula to calculate electrical current is Current = Voltage / Resistance. Electrical current is measured in amperes (A) and represents the flow of electric charge in a circuit. According to Ohm's Law, current is directly proportional to voltage and inversely proportional to resistance. Thus, by dividing the voltage by the resistance, you can determine the current flowing through a circuit.

210. What is the purpose of a junction box in an electrical system?
Answer: To connect electrical wires - A junction box is used to securely connect and protect electrical wires in an electrical system. It provides a safe and enclosed space where wires can be joined together using wire connectors, wire nuts, or other approved methods. Junction boxes help ensure reliable electrical connections, protect the wiring from damage, and facilitate future maintenance or modifications to the electrical system.

211. Which of the following is a unit of electrical resistance?
Answer: Ohm (Ω) - The unit of electrical resistance is the ohm (Ω). It is named after the German physicist Georg Simon Ohm and is represented by the Greek letter omega (Ω). Resistance is a measure of how much an electrical component or material opposes the flow of electric current.

212. What does the term "grounding" refer to in electrical systems?
Answer: Connecting electrical devices to the Earth - Grounding in electrical systems refers to the process of connecting electrical devices, equipment, and structures to the Earth or a grounding electrode system. This

connection serves several purposes, including providing a safe path for electrical faults, reducing electrical noise, and ensuring the effectiveness of overcurrent protection devices.

213. Which of the following symbols represents a resistor in an electrical circuit diagram?
Answer: A zigzag line - In electrical circuit diagrams, a zigzag line is commonly used to represent a resistor. A resistor is an electronic component that restricts the flow of electric current, dissipating electrical energy in the form of heat. It is often used to control the amount of current or voltage in a circuit.

214. What is the purpose of a transformer in an electrical system?
Answer: To step up or step down voltage - A transformer is an electrical device used to transfer electrical energy between two or more circuits through electromagnetic induction. Its primary function is to step up (increase) or step down (decrease) the voltage levels between the input and output circuits. Transformers are commonly used in power distribution systems to adjust voltage levels for efficient transmission and utilization.

215. Which of the following is an example of a renewable energy source used in electrical generation?
Answer: Solar power - Solar power is an example of a renewable energy source used in electrical generation. It harnesses the energy from sunlight and converts it into electricity using solar panels or photovoltaic cells. Solar power is considered renewable because it relies on an abundant and inexhaustible source—the sun—and does not deplete natural resources or release harmful emissions during operation.

216. What is the purpose of a three-way switch in a lighting circuit?
Answer: To control the on/off operation of a light fixture from two different locations - A three-way switch is used in a lighting circuit to control the on/off operation of a light fixture from two different locations. It allows the user to turn the light on or off from either switch position, providing convenience and flexibility. Three-way switches are commonly used in stairways, hallways, and rooms with multiple entrances.

217. Which of the following tools is commonly used to measure electrical voltage?
Answer: Digital multimeter - A digital multimeter (DMM) is a commonly used tool to measure electrical voltage. It combines various functions, including voltage, current, and resistance measurement, in a single device. With its digital display, it provides accurate readings of voltage levels in electrical circuits. Electricians and technicians rely on DMMs for troubleshooting, testing, and maintenance tasks.

218. What is the purpose of a disconnect switch in an electrical system?
Answer: To provide a means of isolating electrical equipment from the power source - A disconnect switch is used to isolate electrical equipment from its power source. It allows for a safe means of disconnecting the electrical supply to specific equipment or sections of the electrical system. This isolation is crucial during maintenance, repairs, or emergency situations. Disconnect switches ensure that the equipment is de-energized, reducing the risk of electrical hazards.

219. In an electrical system, what is the purpose of a ground wire?
Answer: To provide a path for fault currents and protect against electrical shock - The ground wire in an electrical system serves as a safety measure by providing a path for fault currents. It helps redirect these currents into the ground, away from people and equipment, preventing electrical shock and reducing the risk of electrical hazards. Proper grounding is essential for electrical safety and is achieved by connecting the ground wire to the grounding system.

220. Which of the following is a unit of electrical power?
Answer: Watt (W) - The watt (W) is the unit of electrical power. It measures the rate at which electrical energy is consumed or produced. Power represents the amount of work done per unit of time and is calculated by multiplying voltage (V) by current (I). The watt is widely used to quantify the power rating of electrical devices and to assess the energy consumption or generation of electrical systems.

221. What is the purpose of a junction box in an electrical system?
Answer: To safely enclose and protect electrical connections - A junction box is used to safely enclose and protect electrical connections in an electrical system. It provides a secure housing for wire connections, ensuring proper insulation and reducing the risk of accidental contact with live wires. Junction boxes also prevent damage to the wires from environmental factors or physical stress. They are essential for maintaining electrical safety and facilitating future maintenance or modifications.

222. Which of the following is a key function of a circuit breaker in an electrical system?
Answer: To protect against overcurrent and prevent damage to electrical circuits - A circuit breaker is designed to protect electrical circuits from overcurrent conditions. When the current exceeds a safe limit, the circuit breaker automatically interrupts the flow of electricity, preventing damage to electrical equipment and wiring. By quickly

disrupting the circuit, circuit breakers mitigate fire hazards, equipment damage, and electrical failures caused by excessive current.

223. What is the purpose of a ground fault circuit interrupter (GFCI) outlet?
Answer: To provide protection against electrical shock - A ground fault circuit interrupter (GFCI) outlet is designed to protect against electrical shock. It constantly monitors the current flowing in the circuit and can quickly interrupt the circuit if it detects any imbalance or leakage. GFCI outlets are commonly installed in areas where water is present, such as bathrooms, kitchens, and outdoor locations, to reduce the risk of electric shock in potentially hazardous situations.

224. Which of the following is a safety precaution when working with electrical systems?
Answer: Using personal protective equipment (PPE) - When working with electrical systems, wearing appropriate personal protective equipment (PPE) is essential for safety. PPE includes items such as insulated gloves, safety glasses, and flame-resistant clothing. PPE helps protect against electric shock, burns, and other potential hazards. It is crucial to follow safety guidelines and use the recommended PPE to minimize the risk of electrical accidents or injuries.

225. What is the purpose of a transformer in an electrical system?
Answer: To step up or step down voltage levels for efficient power transmission - A transformer is used in electrical systems to step up or step down voltage levels. It allows for efficient transmission of electrical power by adjusting the voltage to an appropriate level for distribution. Transformers are crucial in power generation stations, substations, and electrical grids to match voltage requirements and minimize power losses over long distances.

226. Which of the following is a safety hazard associated with electrical systems?
Answer: Exposed live wires - Exposed live wires pose a significant safety hazard in electrical systems. Coming into contact with live wires can result in electrical shock, burns, and even fatalities. It is crucial to ensure that all wires are properly insulated and protected within appropriate conduits or enclosures. Any exposed live wires should be immediately addressed and repaired to eliminate the risk of electrical accidents.

227. What is the purpose of a capacitor in an electrical circuit?
Answer: To store and release electrical energy - A capacitor is an electronic component used in electrical circuits to store and release electrical energy. It consists of two conductive plates separated by an insulating material known as a dielectric. When a voltage is applied, the capacitor charges, storing electrical energy. It can then release this energy back into the circuit when needed, often to smooth out voltage fluctuations or provide a temporary power boost.

228. Which of the following is a safety precaution when working with electrical equipment?
Answer: Using lockout/tagout procedures - Lockout/tagout procedures are essential safety precautions when working with electrical equipment. These procedures involve de-energizing the equipment, isolating it from the power source, and securing it with locks or tags to prevent accidental re-energization. Lockout/tagout ensures that the equipment is safely turned off and cannot be operated during maintenance or repair work, protecting workers from electrical hazards.

229. What is the purpose of a circuit overload protection device?
Answer: To prevent excessive current flow - Circuit overload protection devices, such as fuses or circuit breakers, are installed in electrical systems to prevent excessive current flow. They detect when the current exceeds the rated capacity of the circuit and quickly interrupt the flow of current to protect the wiring and equipment from damage or overheating. These devices help maintain the safety and integrity of the electrical system.

230. Which of the following is the correct formula to calculate electrical power?
Answer: Power = Voltage x Current - The formula to calculate electrical power is Power = Voltage x Current. Power is measured in watts (W) and represents the rate at which electrical energy is consumed or produced. By multiplying the voltage (V) by the current (I), you can determine the amount of power being transferred or consumed in an electrical circuit.

231. What is the purpose of a ground fault circuit interrupter (GFCI) outlet?
Answer: To prevent electrical shock - A ground fault circuit interrupter (GFCI) outlet is designed to detect imbalances in electrical current and quickly shut off power to prevent electrical shock. It provides protection against ground faults, which occur when current flows along an unintended path, such as through a person or water. GFCI outlets are commonly used in areas where water is present, as they reduce the risk of electric shock in potentially hazardous situations.

232. Which of the following is a safety hazard associated with electrical systems?
Answer: Inadequate grounding - Inadequate grounding in electrical systems can pose a safety hazard. Proper grounding helps redirect fault currents and provides a safe path for electrical faults. Inadequate grounding can result

in electrical shock hazards, equipment malfunctions, or failure of protective devices. Ensuring proper grounding practices and regularly inspecting and maintaining grounding systems are crucial for electrical safety.

233. What is the purpose of a junction box in an electrical system?
Answer: To safely enclose electrical connections and protect against damage - A junction box is used in electrical systems to safely enclose electrical connections and protect them from damage. It provides a secure and insulated environment for joining wires using wire connectors or other approved methods. Junction boxes also help prevent accidental contact with live wires, reducing the risk of electrical shock or short circuits.

234. Which of the following is a unit of electrical power?
Answer: Watt (W) - The unit of electrical power is the watt (W). Power represents the rate of energy transfer or consumption in an electrical circuit. It is calculated by multiplying the voltage (V) by the current (I) and is measured in watts. The watt is the standard unit for quantifying electrical power in various applications, such as appliances, lighting, and electronic devices.

235. What is the purpose of a circuit breaker in an electrical system?
Answer: To protect against overcurrent and prevent damage to electrical circuits - A circuit breaker is a protective device installed in electrical systems to protect against overcurrent conditions. When the current exceeds a safe limit, the circuit breaker automatically interrupts the flow of electricity, preventing damage to electrical equipment and wiring. Circuit breakers play a crucial role in preventing electrical fires, equipment failures, and hazards caused by excessive current flow.

236. Which of the following is a safety precaution when working with electrical equipment?
Answer: Using insulated tools - Using insulated tools is an important safety precaution when working with electrical equipment. Insulated tools have handles or coatings made from non-conductive materials, such as rubber or plastic, which provide a barrier between the user and any live electrical parts. Insulated tools help prevent accidental contact with energized240. Which of the following is a safety precaution when working with electrical equipment?
Answer: Using insulated tools - Using insulated tools is an important safety precaution when working with electrical equipment. Insulated tools have handles or coatings made from non-conductive materials, such as rubber or plastic, which provide a barrier between the user and any live electrical parts. Insulated tools help prevent accidental contact with energized components, reducing the risk of electrical shock or injury.

237. What is the purpose of a disconnect switch in an electrical system?
Answer: To provide a means of isolating electrical equipment from the power source - A disconnect switch is used to isolate electrical equipment from its power source. It allows for a safe means of disconnecting the electrical supply to specific equipment or sections of the electrical system. This isolation is crucial during maintenance, repairs, or emergency situations. Disconnect switches ensure that the equipment is de-energized, reducing the risk of electrical hazards.

238. In an electrical system, what is the purpose of a ground wire?
Answer: To provide a path for fault currents and protect against electrical shock - The ground wire in an electrical system serves as a safety measure by providing a path for fault currents. It helps redirect these currents into the ground, away from people and equipment, preventing electrical shock and reducing the risk of electrical hazards. Proper grounding is essential for electrical safety and is achieved by connecting the ground wire to the grounding system.

239. Which of the following is a unit of electrical power?
Answer: Watt (W) - The unit of electrical power is the watt (W). It measures the rate at which electrical energy is consumed or produced. Power represents the amount of work done per unit of time and is calculated by multiplying voltage (V) by current (I). The watt is widely used to quantify the power rating of electrical devices and to assess the energy consumption or generation of electrical systems.

240. What does the term "grounding" refer to in electrical systems?
Answer: Connecting electrical devices to the Earth - Grounding in electrical systems refers to the process of connecting electrical devices, equipment, and structures to the Earth or a grounding electrode system. This connection serves several purposes, including providing a safe path for electrical faults, reducing electrical noise, and ensuring the effectiveness of overcurrent protection devices. Grounding helps to prevent electric shock and protects both people and equipment from electrical hazards.

241. Which of the following symbols represents a resistor in an electrical circuit diagram?
Answer: A zigzag line - In electrical circuit diagrams, a zigzag line is commonly used to represent a resistor. A resistor is an electronic component that restricts the flow of electric current, dissipating electrical energy in the form

of heat. It is often used to control the amount of current or voltage in a circuit and is an essential component in many electrical applications.

242. What is the purpose of a transformer in an electrical system?
Answer: To step up or step down voltage levels for efficient power transmission - A transformer is used in electrical systems to step up or step down voltage levels. It allows for efficient transmission of electrical power by adjusting the voltage to an appropriate level for distribution. Transformers are crucial in power generation stations, substations, and electrical grids to match voltage requirements and minimize power losses over long distances.

243. Which of the following is an example of a renewable energy source used in electrical generation?
Answer: Wind power - Wind power is an example of a renewable energy source used in electrical generation. It harnesses the kinetic energy of wind and converts it into electricity using wind turbines. Wind power is considered renewable because it relies on a constantly replenished source—the wind—and does not deplete natural resources or produce harmful emissions during operation.

244. What is the purpose of a three-way switch in a lighting circuit?
Answer: To control the on/off operation of a light fixture from two different locations - A three-way switch is used in a lighting circuit to control the on/off operation of a light fixture from two different locations. It allows the user to turn the light on or off from either switch position, providing convenience and flexibility. Three-way switches are commonly used in rooms with multiple entrances or where controlling the lighting from different points is desired.

245. Which of the following is a safety hazard associated with electrical systems?
Answer: Inadequate insulation - Inadequate insulation in electrical systems can pose a safety hazard. Proper insulation helps prevent current leakage, maintains the integrity of wiring, and ensures that electricity flows along intended paths. Inadequate insulation can lead to electrical shocks, short circuits, or fires. Regular inspection and maintenance of insulation are necessary to maintain a safe electrical system.

246. Which of the following tools is commonly used to measure electrical voltage?
Answer: Digital multimeter - A digital multimeter (DMM) is a commonly used tool to measure electrical voltage. It combines various functions, including voltage, current, and resistance measurement, in a single device. With its digital display, it provides accurate readings of voltage levels in electrical circuits. Electricians and technicians rely on DMMs for troubleshooting, testing, and maintenance

247. What is the purpose of a disconnect switch in an electrical system?
Answer: To provide a means of isolating electrical equipment from the power source - A disconnect switch is used to isolate electrical equipment from its power source. It allows for a safe means of disconnecting the electrical supply to specific equipment or sections of the electrical system for maintenance, repairs, or in case of emergencies. The disconnect switch ensures that the equipment is de-energized, reducing the risk of electrical hazards.

248. In an electrical system, what is the purpose of a ground wire?
Answer: To provide a path for fault currents and protect against electrical shock - The ground wire in an electrical system serves as a safety measure by providing a path for fault currents. It helps redirect these currents into the ground, away from people and equipment, preventing electrical shock and reducing the risk of electrical hazards. Proper grounding is essential for electrical safety and is achieved by connecting the ground wire to the grounding system.

249. Which of the following is a safety precaution when working with electrical equipment?
Answer: Using personal protective equipment (PPE) - When working with electrical equipment, it is important to use appropriate personal protective equipment (PPE). PPE includes items such as insulated gloves, safety glasses, and flame-resistant clothing. These protective measures help minimize the risk of electric shock, burns, and other potential hazards. It is crucial to follow safety guidelines and use the recommended PPE to ensure personal safety while working with electrical equipment.

250. What is the purpose of a circuit overload protection device?
Answer: To prevent excessive current flow - Circuit overload protection devices, such as fuses or circuit breakers, are installed in electrical systems to prevent excessive current flow. They detect when the current exceeds the rated capacity of the circuit and quickly interrupt the flow of current to protect the wiring and equipment from damage or overheating. These devices help maintain the safety and integrity of the electrical system.

251. Which of the following is the correct formula to calculate electrical power?
Answer: Power = Voltage x Current - The formula to calculate electrical power is Power = Voltage x Current. Power is measured in watts (W) and represents the rate at which electrical energy is consumed or produced. By

multiplying the voltage (V) by the current (I), you can determine the amount of power being transferred or consumed in an electrical circuit.

EXTRA CONTENTS

Flashcards

These digital flashcards have been designed to consolidate your knowledge and help you pass the journeyman electrician exam.

Audiobook

We are pleased to offer an additional bonus: a complimentary audiobook version. This allows you to reinforce your learning through audio, making it easier to study on the go and ensure you are fully prepared for the journeyman electrician exam.

Electrical Load Calculation

As a third bonus we include an exclusive guide on Electrical Load Calculation. This valuable resource is designed to deepen your understanding of load calculations, crucial for mastering the complexities of electrical systems.

Study Plan

This plan is tailored to guide you in your preparation, ensuring that you cover all the topics you need in a systematic and efficient way for the exam.

Guide to registering for the electrician exam

We provide a comprehensive guide on how to register for the electrician exam. This guide simplifies the registration process, helping you ensure that everything is ready for exam day without any problems.

Playlist Bonus Video

This playlist includes a series of instructional videos that complement the material in the book, providing you with visual learning tools to enhance your understanding and readiness for the exam.

Access to Professional Community and Job Network

As an additional bonus you gain access to professional community. This platform is ideal for networking, sharing insights, and discussing challenges with fellow electricians.
Moreover, this community also serves as a valuable resource for job searching within the electrical industry, providing you with opportunities to advance your career by connecting with potential employers.

Next Steps After Securing Employment

This guide offers practical advice on how to effectively transition into your new role and set yourself up for long-term success.

Scan the QR CODE in the next page

Scan the QR CODE below

https://journeymanelectricianexamprep.newpublishingagency.com/step-1-page-2176-8651

Made in the USA
Thornton, CO
09/30/24 19:11:52

207e490a-e171-4391-b8d0-8afd9107f122R01